THE ROAD RUNS WEST

# THE ROAD RUNS WEST

**A Century Along the
Bella Coola/Chilcotin Road**

## Diana French

HARBOUR PUBLISHING

HARBOUR PUBLISHING
P.O. Box 219
Madeira Park, BC Canada V0N 2H0

Published with the assistance of the Canada Council and the
Government of British Columbia, Tourism and Ministry
Responsible for Culture, Cultural Services Branch.

Cover painting by Diana Durrand
Cover design by Mary White
Page design and composition by Vancouver Desktop
Publishing Centre
Printed and bound in Canada

Photographs credited BCARS are from BC Archives and
Records Service, Victoria, BC. Photographs not credited are
from private collections.

**Canadian Cataloguing in Publication Data**

French, Diana.
   The road runs west
   ISBN 1-55017-114-3
   1. Bella Coola/Chilcotin Road (B.C.)—History. I. Title.
FC3826.9.C5F73 1994      388.1'09711      C94—910734-4
F1088.F73 1994

To my husband, Bob

# Contents

# Introduction

British Columbia has three routes to tidewater: the Trans-Canada Highway, which follows the Fraser River out of the mountains down to Vancouver; Highway 16, the Yellowhead, which cuts across the geographic centre of the province to Prince Rupert; and the third route, Highway 20, the Bella Coola/Chilcotin Road.

Little known and little travelled, Highway 20 is three hundred miles of back road to nowhere much. It leaves the bustling little city of Williams Lake in the heart of the Cariboo and heads west, crossing the Fraser River to climb to the vast Chilcotin plateau. It snakes through the sparsely populated rolling grasslands and jackpine forests before swooping over the Coast Mountain range to the tiny coastal village of Bella Coola.

The road had its beginnings in the Cariboo Gold Rush in the early 1860s. For decades there were dreams and schemes and surveys for trails, roads and railways across the Chilcotin to tidewater. In 1864, an attempt to build a road triggered BC's only Indian "war." The first transcontinental railway almost went through Chilcotin in the 1880s. There were other near misses, but through it all the road just grew, a do-it-yourself project that followed the settlers with little interference from surveyors or engineers.

The earliest settlers were attracted to the Chilcotin by the Cariboo gold rush in the late 1850s. Their farms were on the east edge of the plateau in the Riske Creek area. The next batch, mostly young men, came in the 1880s and 1890s. They ventured farther west, chopping a road through the jackpines as far as Tatla Lake. In 1898 a colony of Norwegians settled in Bella Coola. They built a road up the valley but the mountains blocked access to Chilcotin. A few colonists climbed the mountain to join the handful of First Nations residents and bachelor settlers who lived in the wild west Chilcotin, but they weren't much interested in roads. It wasn't until the late 1940s that the Chilcotin Road was passable in all weather—most of the time—as far as Anahim Lake. There was still a ninety-mile gap between Anahim and Bella Coola.

Bella Coola colonists waited half a century for a road out of the valley. They petitioned and complained in vain. In the early 1950s their children and grandchildren finally took matters into their own hands and blasted their way over the mountains to Anahim Lake, creating BC's third highway to the Pacific.

It wasn't much of a highway. For the next quarter century the Bella Coola/Chilcotin was the longest, worst road in the province. It still is the loneliest. Only half of it is paved. There isn't one supermarket or fast-food chain outlet anywhere along it. The service stations and restaurants close at night. The population of places marked on the map varies from one (Towdystan) to 400 (Anahim Lake). There are more people in the forty-mile-long Bella Coola Valley than there are in the entire Chilcotin.

I first met the Chilcotin Road in 1951 when I went to teach in a one-room school in Chezacut, thirty miles off the main road in the heart of Chilcotin. The only way to get from Williams Lake to Chezacut then, and still, was to walk, ride or drive. There is no public transportation. My brother and father delivered me from the coast in an Austin A7. By the time we got to Alexis Creek we thought we were riding inside a vacuum cleaner. When my dad noticed notched trees on the Chezacut Road he was convinced they were necessary to reassure motorists they were really on a road.

It used to be a tradition—almost a rule—that teachers who came to the Cariboo married and stayed, and that's what I did. I married a cowboy, Bob French, the son of Chilcotin pioneers. He later went to work "on the road" for Public Works. For several years his job was to grade, and sometimes build, roads between Alexis Creek and Anahim Lake in the summertime, and snow-plough them in the winter. We lived at Alexis Creek, but the crews camped on the road, and in the summer I went too. Bob's swampers put up with me—and at one point four small children—as long as I cooked for them and moved camp. I spent a lot of time on the road itself, either hauling stuff from cabin to cabin or waiting for the grader to move the trailer. I became personally acquainted with every pothole in the road, but I also got to know many of the people who appear in this book.

In 1962 we moved to Anahim Lake, where I was introduced to the Bella Coola connection. That was a different Chilcotin, the real frontier. We subsequently left Chilcotin (although Williams Lake isn't far away), but our ties are still there, and when I worked with the local newspaper I tended to focus west of the Fraser.

In the late 1970s the road started to change, certainly for the better for everyone who had to use it. By the mid-1980s it was very different from the road I knew, and nothing at all like the pre-1950s road. When Liz Robertson, a friend who has made more than a few trips on the Chilcotin Road, suggested I write a book about it, it sounded like a good idea. Others thought so as well. So many people helped with the research, and took time to dig out old pictures, that I feel the end result is a community effort, not just mine.

# CHAPTER 1

# Unfriendly Territory

*It is hard to find language to express in adequate terms the utter vileness of the trails in the Cariboo. Slippery, precipitous ascents and descents, fallen logs, overhanging branches, roots, rocks, swamps, turbid pools and miles of deep mud.*
—Lt. Spencer H. Palmer, Royal Engineers, 1862

Old-timers claim anyone who crosses the Fraser River to Chilcotin country stays there. That isn't true. More settlers left than ever stayed. The relentless winds that snapped and snarled across the frozen land in winter and the silent killer cold chased some people away. Others couldn't stand the infamous Chilcotin mud, the mosquitoes, the distances, the silences, the loneliness.

The people who stayed were free spirits. They didn't want to tame the land, they just wanted to be there, so they learned to live with the peculiar rhythm of the Chilcotin drummer who is always a quarter beat out of sync with the rest of the world. They learned the Chilcotin Two-Step—two steps forward, two steps back. They learned Chilcotin time, which is always later.

They learned it really doesn't matter if you put off what you should do today until tomorrow. Or next week. Or next year. Time does wait in Chilcotin, as often as not.

The people who stayed probably saw Chilcotin first in fall, when she is in all her glory. Chilcotin seasons are as blurred as her boundaries, but roughly speaking there are seven of them, sometimes eight. They don't have calendar dates, they come and go as they please, except for fall. Fall knows its place. It comes in September, early September in west Chilcotin, a week or so later in the valley. The sky is higher and wider and bluer in fall, the air is crisper. The countryside explodes into colour—purple hills, bursts of yellow, orange and russet leaves. The grass is goldy-green in the swampy areas, topaz and toast-coloured on the hills and plains. The Chilcotin River—the Tsilhqot'ins called it the "young man's river"—is always milky green, a bright counterpoint to every season's colour scheme.

Indian Summer comes next, taking a last shot with burning days and bringing hints of what is to come with ever cooler nights. October winds strip the leaves, and the wildly coloured landscape gently fades to sepia, like an old photograph.

Freeze-up should come next but sometimes another season shoves in between, a rainy misery known as a Wet Fall. Wet Falls come any time from September to December, can last for weeks or months, and have little to recommend them. The rain drizzles away every vestige of colour, leaving the land drab, depressing and muddy. Freeze-up creeps in quietly, tightening and glazing the ground, gelling the lakes and ponds. Winter pounces. It holds Chilcotin in its icy claws for endless months, but now and then a Chinook wind comes along to whisper "spring" and the winter sun sends promises. Winter is routed by breakup. Breakup is aptly named because the land literally breaks up and brims over. As the snow disappears and the ice rots, the ground thaws, turning into several different kinds of mud, all of them unpleasant. If the sun shines often enough and hot enough, and if it doesn't rain, breakup is over quickly, but if conditions are right low spots can stay soggy until July.

The land looks moth-eaten and bedraggled until spring brings its thousand shades of green. Spring comes like a polar bear and leaves like a bunny, or vice versa, and dithers about in between, but eventually the wildflowers bloom, the birds come back, and all is right with the world, except for the bugs. Sometimes birds return too soon and freeze but the bugs have impeccable timing. The first sign of real spring is mosquitoes. Hordes of mosquitoes. One relation or another stays all summer. When a Chilcotin summer is hot, it is very, very hot, and when it is wet, it is horrid. A good summer has enough hot days to get the hay up and enough wet days to settle the dust, but every kind of summer is too short.

In the beginning, two tribes of people lived in the vast Chilcotin plateau. Canyon Shuswaps claimed the high rolling grasslands above the Fraser River and the lower Chilcotin River valley.

Tsilhqot'in territory began about forty miles west of the river mouth, where the valley widens. It extended south to Chilko Lake and the Coast Mountains, west to the Salmon (now the Dean) River. Nankutlun, the largest Tsilhqot'in village, was in the far west, at what is now Anahim Lake. Smaller groups or bands lived around Tatla Lake, Puntzi, Chezacut and Alexis Creek. Some were south of the Chilcotin River, now known as the Stone and Nemiah areas. The land climbs as it goes west, reaching four thousand feet above sea level at Anahim Lake. The climate varies with the altitude. Winter comes sooner and leaves later in the western highlands where summer is over in the blink of an eye, but forty degrees below zero in the valley is just as cold as forty below on the plains.

Nobody knows how many Tsilhqot'ins there were before the white men came, maybe a thousand, maybe many times more. They roamed a land as big as Holland and Belgium combined. They weren't a wealthy tribe, the harsh land saw to that. They were hemmed in by Shuswaps on the east, Carriers on the north, numerous coastal tribes and the mountains on the south. They feuded with most of their neighbours except for the Nuxalk who lived at the head of North Bentinck Arm, in the lush coastal valley just over the mountains from Nankutlun. Tsilhqot'in prosperity depended on the salmon that battled their way up the rivers every summer to spawn. When the salmon run was heavy, the people had enough food to brave the winter at home. When the salmon failed, many Tsilhqot'ins wintered at the coast where the living was easier. In the other seasons they moved from place to place, hunting, fishing and gathering roots and berries.

Fur trader/explorer Alexander Mackenzie skirted Tsilhqot'in territory on his epic journey across Canada in 1793, but the first white men of record in the area were the North West Company's fur traders. In 1821 the Hudson's Bay Company took over the NWC's territory in what was then called New Caledonia. One of the establishments was the newly built Fort Alexandria on the banks of the Fraser River—about halfway between the modern cities of Quesnel and Williams Lake. Few Tsilhqot'ins went there. In 1829 the HBC decided to tap their market by building a trading post at the junction of the Chilcotin and Chilko Rivers, a four-day trip from Fort Alexandria. It must have been Company men who chose the spelling "Chilcotin" for the people, the river and the plateau. The company didn't have high hopes for Fort Chilcotin. "That part of New Caledonia called Chill Cotten country was settled last fall. Little is expected of it," Jon Tod, the HBC Post Master at Fort McLeod, reported in 1830.

Fort Chilcotin, a few rough log buildings thrown together on a gently sloping, juniper-studded sidehill, opened in October, in a wet fall. It was abandoned the next spring. The Tsilhqot'ins burned the buildings but the Company tried again in 1831. They manned the fort for a few months each year, but they couldn't make a go of it. Fort Chilcotin was costly to operate and it was so far off the beaten track Company men didn't care to stay there. The Tsilhqot'ins were seldom around, and when they were they wouldn't always trade. They liked to wear their furs, they could get whatever they wanted in the way of white men's goods from their coastal neighbours, and they didn't like the Company's intrusion in their territory anyway. In 1836 the Company moved Fort Alexandria to the west side of the Fraser, and closed Fort Chilcotin because the Tsilhqot'ins and Fraser Lake Carriers were warring. In 1842 Chief Factor Donald McLean took charge of what he called the "paltry establishment." McLean was an arrogant bully who didn't even try to hide his disdain for the Natives. During his career with the Company he was responsible for nineteen Indian deaths. His motto was kill first, investigate later. He said Fort Chilcotin "wasn't worth risking the lives of people placed there who are little better than slaves to the Indians being unable to keep them in check." The Company abandoned the fort permanently in 1844, thereby establishing Chilcotin country's reputation as a worthless and undesirable place. It had to be, if the mighty Hudson's Bay Company gave up on it. The Company did establish a fort at Kluskus, a neighbouring Southern Carrier village, but that too was short-lived.

Chilcotin was left alone for the next dozen years and might have been for dozens more if it hadn't been for the discovery of gold in what became known as the Cariboo in the late 1850s. Prospectors poured into the country. Most followed the trail north from Yale to the Cariboo, but Bella Coola was an alternative point of entry for those who climbed the mountains, trekked across the Chilcotin plateau to Fort Alexandria, crossed the Fraser there, and went on to the gold fields. In the early 1860s there were as many as four ships at a time anchored at Bella Coola. Peter Barron, one of the first Cariboo storekeepers, gets credit for roughing out the trail across the plateau. He followed established Tsilhqot'in trails, detouring to find feed and water for horses. William Manning, the first known settler in Chilcotin, established a farm at Puntzi Lake. Here he made hay from swamp meadow grass and provided a stopping place for packtrains and travellers. Although Manning's farm was on a traditional Tsilhqot'in gathering place, he got along well with his neighbours, and had a Native wife. The Tsilhqot'ins were blamed for killing a number of intruders, but nothing stopped the tide of gold-seekers.

Farmers followed the miners, and in the mid-1860s, a half dozen or so settled west of the Fraser River on the eastern edge of the Chilcotin plateau. Among them were Thomas Meldrum, who took up land in a valley south of Alexandria; L. W. Riskie, who settled near the Chilcotin River twelve miles west of the Fraser; and the two Withrow brothers, who farmed just west of Riskie. Two American brothers, Thaddeus and Jerome Harper, established the Gang Ranch south of the Chilcotin River in the early 1860s. It was, and still is, one of BC's largest cattle ranches. It was named for the plough used to till the land. The Harpers specialized in beef, but the Chilcotin ranchers raised pigs, grew vegetables and wheat, built a grist mill to make flour, and packed their products along a trail high above the Fraser River to Fort Alexandria.

The Tsilhqot'ins had so little direct contact with the fur traders they were relatively free from alcohol and disease. Although they had even less contact with the miners, that contact brought disaster. In the spring of 1862, a smallpox epidemic started in Victoria and spread like wildfire through the vulnerable Native population. It raged through Bella Coola and Chilcotin, turning villages into stinking graveyards when there was no one left to bury the dead. A few hundred Tsilhqot'ins, scarred in body and savaged in spirit, survived the plague. The Canyon Shuswaps died or fled, leaving their villages deserted. Old Guichon, a Tsilhqot'in who lived near Tatla Lake, was one of the survivors. When the first white settlers reached Tatla Lake in the 1880s, he told them that when he was young, he climbed to the top of a nearby bald mountain where he could see, in the distance, the smoke from many campfires. He said he got sick and went away. When he came back, he climbed the mountain but he could see no campfires. They were gone, like the people in the villages.

The miners weren't interested in dying Indians. Neither was the colonial government. Their only interest was better access to Cariboo gold. A number of entrepreneurs scouted Chilcotin and made proposals for pack trails and wagon roads. Several contracts were signed, and the Fraser River Road Company Ltd. actually built a road and packed freight on it, but they never got a boat contract to Bella Coola and nothing came of any of the schemes. In 1862 Governor James Douglas decided the colonial government would build a road to the goldfields and in July, in the middle of the smallpox epidemic, he sent Lieutenant Spencer H. Palmer of the Royal Engineers to Bella Coola to survey the Barron Trail. Palmer recommended against the route (which became known as the Palmer Trail) because of the difficulty in getting out of the valley and up the mountain. There were two major obstacles: the Great Slide, over 1100 feet of disintegrated rock clinging to a mountainside, and the Precipice, a mountain of basaltic rock 1350 feet high. The Royal Engineers built the Cariboo Waggon Road along the Fraser River instead, and when it was finished Bella Coola became the back door to the goldfields. Nobody much used it. The town of Soda Creek, which sprouted on the Fraser River south of Fort Alexandria, became the front door to Chilcotin.

The wagon road gave New Westminster merchants a direct route to the Cariboo, but it did nothing for Victoria, so when entrepreneur Alfred Waddington came up with a scheme to build a road from the head of Bute Inlet to the goldfields, island merchants leaped like trout to the bait. Waddington, an Englishman who had had various business interests in the US, was lured to Victoria by the gold rush. He bought land, started a few businesses and flirted briefly with politics. He was in his fifties when he and a few other Victoria investors formed the Bute Inlet Wagon Road Co. Ltd. and received a charter to build a toll road. He envisioned a steamship connection to Victoria and thought his road would open "the vast fertile lands" of Chilcotin to settlers. He had a trail roughed out in 1863 and the next spring he sent seventeen men to Bute Inlet to start building the road through the Homathko River Canyon. The Waddington route joined the Palmer Trail at Puntzi, and Alex McDonald, a packer on the Palmer Trail, was hired to take supplies from Bella Coola to Puntzi.

James Douglas retired early in 1864, and Frederick Seymour was appointed Governor of British Columbia. He hardly had his chair warmed when all hell broke loose in Chilcotin. A small band of Tsilhqot'ins attacked the Waddington crew in the Homathko, killing all but three men, who escaped. The warriors headed west, just as McDonald left Bella Coola with seven men and a forty-two-horse packtrain. The Tsilhqot'ins stopped at Puntzi, killed William Manning, the lone settler, then ambushed McDonald near Nankutlun. They killed him, two of his packers, and one packer's Tsilhqot'in wife.

When the Homathko survivors reached Victoria and told their story, Governor Seymour thought he had a major uprising on his hands. He directed Magistrate William Cox of Quesnel to organize a posse and head for Chilcotin, while he organized an army to sail to Bella Coola for an overland chase from that direction. When Seymour, Chief Magistrate Chartres Brew, and forty-eight members of the New Westminster Rifle Corps arrived in Bella Coola, they found the sixteen white residents and McDonald's five surviving packers holed up, expecting the worst.

Events move slowly in Chilcotin. The Tsilhqot'ins attacked the Homathko crew on April 19, but Cox and his sixty-five men didn't get to Puntzi until June 12. When Cox found Manning dead and his place ransacked, he built a small garrison and bravely dug in to wait for Seymour. The governor, who had recruited some Nuxalk men, arrived at Puntzi July 6. He promptly launched a futile search of the rugged and unexplored country for the wanted warriors. He never would have found them, but Chief Alexis, a Tsilhqot'in known to Cox, persuaded eight suspects to come in and parley. Instead of negotiating, Cox arrested the men and took them to Quesnel where they stood trial before Judge Matthew Begbie. Five were hanged. Donald McLean, the former Hudson's Bay trader, was the army's only casualty. A volunteer with Cox's posse, he went out on a sortie and was shot, no doubt by a Tsilhqot'in with a good aim and a strong grudge.

Much has been written about this incident. It has been called everything from a massacre to a war. According to all accounts, Waddington's men treated the Natives abominably, refusing to give them food when they worked for it, treating them like dogs, mistreating the women, even threatening another smallpox epidemic. The result was hardly a war, or even an uprising, because few Tsilhqot'ins joined in, but it did have some immediate results. It stopped Waddington's road, although it didn't stop the white invasion. It triggered the government into moving the Tsilhqot'ins east so as to keep an eye on them (by 1870 Nankutlun was deserted). And it gave the Tsilhqot'ins their first experience with clearcut logging. Government forces razed Puntzi's forests for firewood and killed every edible animal they could find. The expedition was costly too: the government spent $80,000 on it.

When Fort Alexandria closed in 1865, independent traders took up the slack. Several set up shop in the lower Chilcotin and Dan Nordberg settled in the valley. It is said Nordberg was a sailor who "jumped" from a fur trading ship in Bella Coola and went to the Nankutlun country to trade with the Tsilhqot'ins, moving east with them in the late 1860s. It was Tom Hance who began the road west into Chilcotin. The palmy days of the gold rush were over when Hance arrived on the Cariboo from the eastern United States in the mid-1860s, so he went to work for Gang Ranch. Intrigued by stories of Chilcotin, he went to see it for himself and liked what he saw. He and a partner, B. F. (Doc) English, another American, established a trading post on what had been the border of Shuswap and Chilcotin territory, about forty miles west of the Fraser, across the Chilcotin River from the Stone Band. Hance chopped out a wagon road above the Chilcotin River—some swore the trail actually hung out over the bank. He and English later extended the road to meet the Soda Creek ferry.

*The Hance boys. Grover Hance (left) was born in 1888 at Soda Creek. The eldest of the Hance children, he was the first white child born in the Chilcotin Valley. Grover married twice but had no children. Percy (right) never married. He was known for his unfailing courtesy. No one ever saw him lose his temper. Grover, Percy and Rene were known as "the Hance boys" even when they were long past being boys. The trio operated a successful guest ranch where "dudes" came from far and near just to meet the brothers.*

*Rene Hance, 1950s. Rene, third and youngest of the Hance sons, was dashing and debonair. He served as coroner for over forty years, a provincial record that may still stand. He ended his working career as a provincial court judge in Williams Lake.*

*Nellie Hance. Nellie, shown below with four friends, was sixteen in 1886 when she married Tom Hance and left her home in Victoria for the wilds of Chilcotin. She travelled by steamboat as far as Yale, then by saddle horse, the only way to get to Hanceville. The young bride rode 300 miles to her new home with her husband's packtrain.*

There was never even a hint of trouble between English, Hance and the Tsilhqot'ins. By all accounts Hance was a charming man, and a fair one, and he must have had a good ear for the Drummer because he stayed in Chilcotin for the rest of his life. The partners sold their furs and bought supplies at Yale, the nearest trading centre, taking packtrains along the southern route through Gang Ranch and joining the Cariboo Waggon Road at Clinton. They packed in millstones for a grist mill and a saw for a sawmill, and they sold their products to the mines in and around Barkerville. The round trip to Yale took over two months so the partners took turns, Hance going one year, English the next. English liked a good horse race and he liked to bet. On one trip he got involved with some traders with like minds and he won $500 from them. Two years later they were waiting for him. They won all his money and the fur money as well. When he got home he gave his share of the business to Hance to settle the debt and moved to lower Chilcotin where he started Deer Park Ranch. He later sold out and moved to Ashcroft, where he became famous in the Cariboo for raising—what else?—race horses.

CHAPTER 2

# Mudpups and Other Adventurers

*There are magnificent grazing plains in Chilcotin, rich valleys, and an abundance of water, pure and clear. Grass is two feet high, there is plenty of hay for winter, and the cattle are in good shape in the spring. There is a promise of minerals in the area, trappers are doing well, settlers are prosperous. Women and children are robust, but there is a lack of church and education facilities. Mosquitoes are a problem on the plains, but isolation is the main obstacle, we are missing a road to get cattle out and supplies in.*
—Chilcotin rancher Mortimer (Old) Drummond, interviewed in the *Vancouver Daily World*, October 1889

As Tom Hance discovered, the easy pickings in the Cariboo goldfields were over by the mid-1860s. Men were still mining, but with machines and equipment that required capital investment. As the gold rush withered, any attraction Chilcotin had for settlers withered with it. The established ranchers, L. W. Riskie, Thomas Meldrum and a few other settlers stayed, but few newcomers were interested in the place until British Columbia joined confederation in 1871, and surveyors began to eye the plateau as a route for the transcontinental railway. As many as 300 men at a time swarmed around Chilcotin surveying seven different routes. One echoed Waddington's dream, following the Homathko River to Bute Inlet; another looked at Bella Coola as a possible terminus. Premier Amor De Cosmos favoured the former route. He thought it would open what he believed to be Chilcotin's "magnificent agricultural land" and bring prosperity to Vancouver Island. He was furious when the Dominion government stopped all land development in the area until reserves were allotted for the use and enjoyment of the Indians. The Cariboo representative to the provincial legislature, George Walkem, wanted the area settled too, but he was leery of the Tsilhqot'ins. He knew about John Salmon, who, in 1871, had taken up land in the Chilcotin valley with two partners. They had fenced ten acres, and were running a hundred head of cattle and ten horses when Chief Alexis told them to leave. Salmon appealed to Cariboo Magistrate Peter O'Reilly, who consulted with rancher L. W. Riskie. Riskie reported the Tsilhqot'ins were interested in learning to farm, but they claimed the land, and let the settlers know it. "We are barely tolerated," Riskie wrote. "If the Indians aren't allowed all the liberties they want, they could order us out. Someone ought to talk to them, people on this side of the river feel very exposed."

O'Reilly "talked to them." He explained confederation, railway surveys and settlers' rights, but his report to the government noted Salmon mistreated some Tsilhqot'ins. Convinced that three men suspected of being involved in the 1864 skirmish were out to kill him, Salmon moved his stock and himself to Soda Creek. The land freeze was lifted in 1873, although the government didn't herd the Tsilhqot'ins onto reserves until the 1880s. O'Reilly doled out the reserves, keeping the best land for settlers. One exception was Chief Anaham's reserve west of Hance's. When Anaham and his band left Nankutlun, they moved to some of the best farmland in the Chilcotin valley. Anaham was more than a match for O'Reilly. He refused to move again and the commissioner didn't care to force the issue.

Settlers did not crowd into Chilcotin when the freeze was lifted. Those who came expected to find lush meadows, clear, pure creeks, streams and lakes flapping with fish, forests teeming with game and fur-bearing animals, and a railway. The land lived up to expectations, but the railway didn't materialize despite the efforts of De Cosmos and Marcus Smith, BC's chief surveyor. Smith was so obsessed with the gloomy grandeur of the Homathko he staked his career on the route, fighting for it to the bitter end. By the time the new settlers realized they weren't even getting a branch line, they were either hooked on the country or stuck there.

Chilcotin was as inhospitable to the newcomers as she had

been to the Hudson's Bay people and the miners, but some of the new wave heard the Drummer. They were adventurers, escapists, eccentrics, or men with no future at home. They wanted to make the first footprints in the snow, to ride with the wind, to stand on top of the world. They wanted to have their own kingdoms and they were stubborn enough to think they could. Among the men who came from older, gentler lands to challenge Chilcotin were Frederick Becher, Hugh Bayliff, Norman Lee, and Alex Graham. Becher was a Northern Irishman who worked for the Hudson's Bay Company at northern posts before finding his way to Chilcotin in the late 1870s. He joined George Dester, an English sea captain who started a trading post near Riskie's after Fort Alexandria closed. When their property was included in Chief Toosey's reserve, the partners moved down country a few miles.

Bayliff was a mudpup, one of the young men from British families who apprenticed on established Cariboo ranches. Under this system the family paid the rancher $500 a year (plus $125 for a horse) to teach the young gentleman the art of ranching in the new country. The pupils were called mudpups because they did the chores—mucking manure from the barns, fencing, and riding in all kinds of weather looking for stray stock. If the lad took to the life, his family helped him get a ranch of his own. Bayliff was the son of a clergyman. Husky and athletic, he rode like a centaur. After serving his time on the well-established Cornwall Ranch at Ashcroft, he worked at the Roper Ranch near Kamloops. In 1885, Roper sent him to deliver some horses to Tom Hance in Chilcotin. Then in his fifties, Hance was still the only settler in the valley on the north side of the river. (His land pre-emption number was 1.) But there was one homesteader, Mike Minton, on the south side of the river, past the Stone Reserve. Hance had extended his fur trade to the Southern Carrier village of Ulkatcho, north of Alexis Creek, on the old Mackenzie Trail. His business affairs took him to Victoria on occasion, and he returned from his 1886 trip with a bride, Nellie Verdier. The petite sixteen-year-old rode from Yale to Hance's. She was the first, and for several years the only, white woman in the Chilcotin valley.

Hance sent young Bayliff off to Ulkatcho with a packtrain of supplies. The well-used trail followed the milky green Chilcotin River west from Hanceville and the valley was a riot of colour. The far hills to the south floated in a soft blue haze, the near hills were clad in dark green evergreens dappled with shimmery yellow aspen. The golden leaves of the cottonwood trees along the riverbank were so bright they seemed to glow with a light of their own. On the north side of the river trail, a high, purplish-brown rimrock ridge ran the length of the valley. It was flanked with bunchgrass-studded sidehills and lush swamp meadows ringed with winey-red willows. The grass was hip-high to a tall horse. Over it all was the bottomless sky, paintbox blue with woolly white clouds in the bright, warm days, velvety black and swarming with stars in the frost-nipped nights. Sleek, fat-bellied deer browsed in the willows, lazy from a summer's overindulgence. Squirrels dashed around loading their larders. Every pond and pothole was busy with ducks and geese getting ready for their trip south.

Bayliff thought he was in paradise after the drought-scorched Kamloops country. He returned the next spring and staked land by the Chilcotin River some thirty miles west of Hance's. Back in Kamloops he recruited a friend, Norman Lee (another English clergyman's son), who was clerking for the Hudson's Bay. They got a hundred head of cattle from Roper on shares, bought some trade goods and set out for Chilcotin on the Gang Ranch route.

They had a terrible trip. They had to swim the animals across the swollen rivers, and the Shuswap riders they hired deserted them when they reached Tsilhqot'in country. Thanks to Bayliff's mudpupping experience and his skill as a rider, they got the herd home intact and settled in. Both men walked easily to the Drummer's beat. Their cattle were prolific and the Tsilhqot'ins traded at their store. Still, their life was not filled with wine and roses. They lived on salmon, bacon, rice, beans and tea. They were often cold, sick, miserable and lonely, but they survived everything Chilcotin threw at them.

When the Canadian Pacific finally got itself across Canada, it brought some benefits to Chilcotin. The railhead at Ashcroft was closer than Yale. It took just over a month to drive cattle there and bring supplies back. Another benefit was less obvious. Many of the laborers imported from China to build the railway were set adrift when the work was done, and some of them found their way to Chilcotin. Everybody who was anybody had a Chinese man to do the cooking, housework, chores, gardening, to mind the children if there were any, and to irrigate the hay fields. These unsung heroes worked from dawn to dark, seven days a week, for whatever "bossy" decided to pay them.

In 1887, Riskie and the Withrows retired after twenty years in Chilcotin, selling out to Mortimer Drummond and F. M. Beaumont. Alex Graham arrived that year too. At eighteen, Graham was a wiry lad with rusty red hair. He was a Scot, but raised in Ireland. He and his buddy Archie Macauley landed in Washington state in 1886 and worked their way north. They stopped at Chilcotin, where Graham went to work for Beaumont while Macauley scouted for land up the valley. Graham sent for his childhood sweetheart Anna Harvey in the fall of 1889. The twenty-year-old, copper-haired teacher crossed the ocean and continent by herself, expecting Alex to meet her at Ashcroft. He didn't, so she took the stagecoach to Soda Creek, a rough, dusty, three-day trip. He wasn't there either, and she waited two days, the only woman at the rough hotel. She had no money left and thought she'd been jilted, but she was simply a victim of Chilcotin time. Whoever picked up the mail didn't get around to giving her letter to Alex in time for him to meet her. When he did arrive, there was the problem of marrying. They took the stage south looking for a preacher—they were in short supply—and were fortunate to find one in Clinton. Back at Soda Creek a week later, Anna crossed the Fraser River in a flat-bottomed ferryboat with a rambunctious ram on its way to Drummond's. Anna was so terrified she vowed she'd never get in a ferry again, and she didn't. On the Chilcotin side, the new couple faced a fifty-mile ride. Anna knew nothing about horses and her voluminous skirt with its fashionable bustle was hardly the outfit for riding astride a stock saddle. Saddle sore and terrified, she wasn't impressed by the splendour of the autumn landscape. The newlyweds stayed one night with Meldrums and the next at Becher House, reaching their one-room, sod-roof log cabin on the third day.

Grover Hance, the first white child born in Chilcotin, arrived in 1888. Tsilhqot'in women assisted Nellie Hance. In 1891 she attended the birth of Graham's daughter Frances, and so became Chilcotin's official midwife. That same year, Hugh Bayliff returned from a trip to England with a bride, the daughter of a prominent London businessman. The Bayliffs spent a few days with the Cornwalls at their gracious Ashcroft home, and that visit can't have prepared Gertrude Bayliff for her little log home in the wilderness. When she arrived there (after riding sidesaddle from Cornwalls) she found Norman Lee, her husband's partner, sitting at the table, mending a nasty gash in his leg with a needle and thread. The cabin had a rough whip-sawn board floor, and Gertrude broke the heels off all her shoes, catching them in the cracks. However, she never lowered her standards. Chilancoh Ranch became another bit of old England on the vast frontier. Mrs. Bayliff set her table with silver, dressed for dinner, and established as high a tone as was possible under the circumstances.

Bayliff and Lee had agreed that when one of them wed, they would toss a coin and the winner would buy out the loser. Lee won the toss, but he couldn't raise the money. Bayliff could, so he got the ranch. Lee bought trader Dan Nordberg's place near Hance's. Nordberg moved to Riske Creek.

Land boundaries in Chilcotin were, and still are, as vague as the sense of time, but generally speaking, the eastern plateau from Meldrums to Withrows, the first area to be settled, was called Chilcoten (with an "e"). The name became official in 1886 when the population warranted a post office with Riskie as postmaster. Fred Becher was postmaster from 1894 until he died in 1936. In 1911 the spelling was changed to Chilcotin ( with an "i"). In 1912 the name of the post office was changed to Riske Creek (with no "i" but still pronounced Riskie). It takes a long time for change to be noticed in Chilcotin and the road was still spelled Chilcoten (with an "e") until the late 1920s. When Tom Hance got a post office and the mail contract in 1889, the eastern end of the valley became known as Hanceville. The mid-valley, Chief Alexis's territory, was called Alexis Creek early on. Two Englishmen named Hewer and Nightingale pre-empted what had been one of Alexis's villages beside a creek, and the name became official when they got a post office in 1893.

*The Lees. Norman (left) and Penrose (E. P.) Lee were brothers, sons of an English clergyman. Norman arrived in Chilcotin in 1887 and later established a ranch and trading post at what is now Hanceville. E. P. came to visit in 1888 and returned in 1895 to start a ranch at Redstone. His sister, Helen Lee, joined him in 1921. Norman married his cousin, Agnes Lee, and toward the end of the century her brother, Thomas Campbell (T. C.), settled in Alexis Creek as well. (BCARS 94624)*

*Bella Coola Road. For the first ten years or so of its existence, the Bella Coola connection was little more than a wagon road through the bush. But to valley residents, hemmed in by the mountains, it really was the Freedom Road. Intended as a road for motor vehicles, it pleased people with saddle horses and wagons just as much because it was such a vast improvement over existing trails.*

*Packhorses on the Precipice. For almost 100 years this mountain of basaltic rock 1350 feet high, and the Great Slide, a mile of disintegrated rock clinging to the mountainside, blocked all attempts to build a road out of the Bella Coola Valley. But the Palmer Trail, used in gold rush days, tackled both obstacles. Until the road was built on the other side of the valley in the 1950s, packers used the precipice trail to transport everything from mowing machines to cookstoves up the hill. What went down was mostly furs. (BCARS 64711)*

*C1 Ranch House, Alexis Creek. Built around 1910 by Alex Graham, the house is no longer there.*

*T. C. Lee's, Alexis Creek. The original Lee home and stopping place still sits beside the road at Alexis Creek, although it is empty. It is one of the oldest buildings in Chilcotin.*

Most of the settlers were British gentry, mainly because land pre-emptors had to be British subjects or take an oath of allegiance to the Crown. The gentry also had the wherewithal to get started. The young Brits established themselves before sending for or fetching wives from the Old Country, but when an American immigrant, Benjamin Franklin, arrived from Washington state in 1886, he brought his wife and two small children with him. Franklin was different, even for Chilcotin. He must have been born under a wandering star because he couldn't stay put. He'd find land he fancied, start a ranch or a store, or both, sometimes with a partner, sometimes not, then he'd move on. When he arrived in Chilcotin, he left his family at Hanceville while he chopped a road to Tatla Lake. Franklin had his eye on land at the west end of the lake, but a Tsilhqot'in named Old Guichon claimed it. Franklin tried to explain about Crown land and pre-emptions, but Guichon said no Queen had ever talked to him. Franklin gave him a caddy of tobacco for the property and established a store there. It's a beautiful spot, wide-open meadow land flanked by rolling, grassy hills on the north, and high snow mountains on the south. For ten years or so, Mrs. Franklin lived at Bute Inlet where there were fifteen or so families. Benny backpacked supplies from there to Tatla through the Homathko canyon.

Benny built roads all over the place, including the one from Norman Lee's to the Gang Ranch. He carried an axe wherever he went and wore a huge black hat that made him look like a wee walking mushroom. He was convinced Chilcotin would flourish once there was a proper way out of the place, and he never gave up looking for one. One of his expeditions was reported in the Victoria *Colonist* in 1892. In March that year, accompanied by two Tsilhqot'ins and travelling by horse, foot and snowshoe, he followed an old trail to Knight Inlet, a 130-mile trip through vicious country. The trio then canoed to Victoria. The trip took three weeks. The *Colonist* reported the Tsilhqot'ins were astonished by Victoria, and Victoria residents were equally astonished by them. Franklin convinced government officials to survey the route, but they found the way blocked by a box canyon.

Alex and Anna Graham joined Archie Macauley at Alexis

Creek in 1893. According to the provincial government's annual report, the 35 miles of road heading west from Riske Creek followed the north side of the Chilcotin River "through what appears to be the best agricultural part of the valley." It was deemed passable for loaded wagons. The report did not mention that the road teetered along the riverbank, nor that it was littered with windfalls and boulders where it wasn't muddy. The Grahams spent almost a week travelling to their new home. Alex rode saddle horse, driving their few head of cattle. Anna and baby Frances jolted along in a buckboard, a part-wagon, part-buggy contraption loaded with all their belongings. The June weather was friendly. It didn't rain. The days were pleasantly warm, the nights cool. The valley was newly green, freckled with bright wildflowers. The only sour note was the mosquitoes. They were a problem on the plains, in the swamps, in the forests—everywhere. They were ecstatic with this rare human feast and Anna was dizzy from swatting them. Frances was swaddled to the eyes but it didn't protect her from the little beasts. The road got worse the farther it went. As well as prying the buckboard out of holes and over big rocks, the Grahams chopped through windfalls and felled a tree or two, building road as they went west, into the sunset.

CHAPTER 3

# The Promised Land

*Many years ago, before the white man even knew this land existed, there was a winter of deep snow, followed by a springtime of heavy rains. It rained for weeks, and the melting snow filled the rivers and streams to overflowing and land everywhere was under water. In the Fraser Valley people took to their canoes to wait out the flood. One night a fierce storm came up and swept the canoes out to sea. One of the big canoes was battered in the storm for many days and when the wind went down, the people saw they were near a peak of land, so they tied the canoe to it with a cedar rope. The rain finally stopped, and the sun shone and when the water finally receded the people saw they were at the top of a high mountain overlooking a beautiful valley, a valley as rich and beautiful as the one they left. They called the valley Bula Kula, the Place of the Broken Anchorage, and they called the mountain Noosgulch. Today the valley is called Bella Coola, the mountain is called Nootsatsum, or Nutsatsum. If you look carefully you can still see, high up on the mountain peak, the mark left by the cedar rope.*
—Nuxalk Legend

While newcomers straggled into Chilcotin, Bella Coola was settled in one fell swoop.

The first white men to see the Bella Coola valley were the fur trader and explorer Alexander Mackenzie, and his party. They passed through the valley in July 1793, on their epic voyage across Canada by land. Captain George Vancouver was exploring the coast at the time and one of his boats missed Mackenzie's party by a few weeks. It is hard to see how anything would have been different had they met, but it was the first in a number of near misses that plagued Bella Coola over the years. There were several thousand Bella Coolas (Nuxalk) then, living in villages strung along the river between tidewater and Shtooiht (Stuie), and for fifty years Bella Coola, at the end of North Bentinck Arm, was a port of call for marine fur traders. If destiny had cast more than sheep's eyes at Bella Coola, the place might have become one of British Columbia's major ports. During the gold rush it nearly did—there were as many as four ships in the bay at once then—but the combination of the Cariboo Waggon Road and the Chilcotin war put an end to that.

There are no plains along this stretch of BC coast. The mountains rise right out of the sea, and the bays and inlets poke through them. Bella Coola is on tidewater, but it is sixty miles from open water. That is probably why the Hudson's Bay Company established Fort McLoughlin near Bella Bella in 1833, but didn't get around to Bella Coola until 1867. Seven years later, a young Englishman named John Clayton walked from Barkerville to Bella Coola, lured perhaps by the railway surveys. He worked at the HBC post and managed it when the postmaster died in 1876. When the Company abandoned the fort ten years later, Clayton bought the property. A combination of bad men, bad booze and bad sick (smallpox) had decimated the Nuxalk. First the marine traders' rotgut booze dulled their will to fish and hunt, and then the miners' smallpox swept through the valley. The little villages along the river vanished into the forests, leaving only ghosts and the winds in the cedars singing requiems. In 1889, government legislation established one large reserve at the mouth of the Bella Coola River. The reserve was on the north side of the river, and Clayton kept land on the south side. When Fort McLoughlin closed he bought it too. He virtually ran both districts, the master of all he surveyed. Fortunately he was a kindly potentate.

Bella Coola might have remained as it was except for B. Fillip Jacobsen. In 1884 this young Norwegian travelled from Vancouver to Alaska, alone, in a small boat, stopping at Indian villages to collect artifacts for a German museum. So fascinated was he by Bella Coola people, he returned the next year and persuaded eight men to go with him on a thirteen-month tour of Europe. The men performed traditional dances in ceremonial costumes and were a big hit wherever they went. Jacobsen was also fascinated with Bella Coola, and when he returned from the tour he talked the provincial government into surveying land in the valley for settlement. He brought in a few settlers himself, and he wrote articles for Norwegian-language publications, describing Bella Coola in glowing terms and comparing it favourably with Norway. His reports caught the eye of Reverend Christian Saugstad, pastor of the Lutheran Free Church at the Augsburg Chapel in Cookston, Minnesota. Saugstad's flock were recent Norwegian immigrants who were suffering hard times economically and spiritually. Crops were poor, markets worse, and there were rifts within the church. Saugstad was seeking greener pastures for his followers. The pastor was an idealist: he was looking for the promised land of milk and honey where his people would be free to worship as they wished, far from all worldly temptations. He found his Utopia in Bella Coola. The colonists' descendants still scratch their heads wondering why he chose an isolated valley in northern BC when he could have settled in the Yakima Valley in central Washington. They think he must have been bedazzled by the mountains or looking through a rainbow.

The BC government welcomed Saugstad with open arms and promised if he could produce twenty families, each with $300 cash, every family would receive 160 acres of land and money would be provided to build a road. There was a catch—the land had to be taken up in a block. Saugstad recruited colonists from the upper Red River Valley of Minnesota and the Dakotas. They elected a committee to govern the new colony, collected money, and drew up a constitution. Colonists were required to be industrious and loyal and they had to furnish satisfactory evidence of good character. One item in the constitution had a far-reaching effect: it prohibited the use of intoxicating drinks in the colony except for sacramental, medical, mechanical and chemical uses. The colonists meant business on this score. One of their number who became tipsy en route was sent back to Minnesota.

There were sixteen white people living in Bella Coola during the height of the gold rush in 1864, and that's how many there were thirty years later when the Norwegians arrived. John Clayton, his wife Elizabeth and their two sons lived in comfortable circumstances in a huge house beside the Bella Coola River. Their home was a bit of old England, the fine Victorian furnishings

*Roadbuilding in Bella Coola, 1899. When the Norwegian settlers arrived in Bella Coola in 1894, they had to clear giant trees and thick underbrush from the land for their homesteads and for the road up the valley. They had no horses or oxen, and their only convenience was a wheelbarrow. (BCARS 74225)*

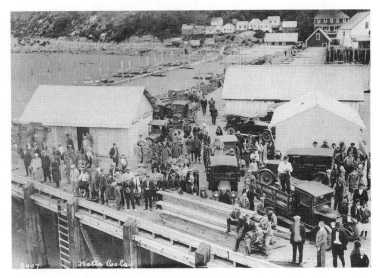

*The Board of Trade visits. Until the road was built to Anahim Lake, the Union Steamship boat was the only dependable link between Bella Coola and the outside world. On Thursday, Boat Day, everyone turned out to greet the steamship and pick up mail and supplies. Occasionally there were special visits. In 1920, members of the Board of Trade from Seattle and Vancouver chartered the steamship* Princess Alice *to visit communities along the coast. Bella Coola had an active Board of Trade at the time, led by Adolph Christensen and Berger Brinildsen. The* Princess Alice *received a royal welcome from valley residents.*

*Bob Pyper's cabin, Chilanko Forks. Chilanko Forks has moved around a bit, depending on who had the post office. Bob Pyper, the first of several provincial policemen to leave the force and remain in Chilcotin, took up storekeeping. Early Hodgson drivers used to stop over at Pyper's, when the old man wasn't too cranky. (BCARS 99860)*

enhanced by Indian artifacts. Clayton had cattle and horses too. Some missionaries were residents, and one of Jacobsen's settlers, retired sea captain Thor Thorsen, his wife and two daughters, lived on the banks of the Smootl River (now Thorsen Creek). Among the bachelors was Thomas Draney, an Irishman who worked at his brother's fish cannery in Namu in summer and wintered in Bella Coola.

The first eighty colonists, mostly men, arrived aboard the sidewheeler *Princess Louise* on October 30. Until they had communal shelters built, they camped in Jacobsen's field beside the Bella Coola River, five or six men to a tent. They drew lots for their property in groups of four so they could help each other clear land and erect buildings. Right away they realized they weren't in the promised land. Coastal communities can be dismal in late autumn; even Rev. Saugstad must have wondered at this choice. The valley was dreary. The mountains loomed over the newcomers, critical of every move. There were no open meadows in the valley. Every inch of land was buried under thick tangles of brambles and bush, windfalls and enormous trees. And then, as if the cold, relentless November drizzle weren't misery enough, the tent camp was treated to a torrential rain and the Bella Coola River decided to defend her territory with a major flood.

The homesites were up the valley, and time had long since stolen the Palmer Trail. When five men thrashed their way up to look at their property, the forest was so gloomy and thick they had to use compasses and signal lights to keep track of each other. Rather than fight their way back, two men tried to navigate the swollen river on a hastily built raft, and they nearly drowned when the current knocked it to bits. Some colonists left in disgust before they even looked at their land; a few went when they saw what they had; others wanted to go but couldn't afford to.

The men grubbed out a trail through the soggy black forest. They had no horses, so they either backpacked up the trail or hired Native people to carry supplies upriver in their spoon-shaped canoes. Winter ice soon put an end to river transport, leaving the colonists with the wearying twelve-mile trek to tidewater. Clayton's store had everything the colonists needed—shovels, axes, buckets, pots and pans, jars of canned fruit and meat, patent medicines, rifles, ammunition. The door to his house was always open, when the colonists came to shop they were always invited in for a meal. Clayton did everything he could to help the new-comers, but he couldn't do anything about the weather which did everything it could to discourage them. After the floods receded, the east wind pushed the arctic air off the mountains into the valley where it howled through the forest to the sea, dropping freezing rain along the way. Then it snowed—wet, gloppy snow. There were days of dreary fog when the entire valley was draped in a dripping blanket of cold mist. Even the good days were cheerless; the winter sun was too weak to climb over the moun-tains and sparkle frost hung thick and chilly on the trees and bushes all day long. The colonists persevered, spurred on by rumours that the government expected them to fail.

The Norwegians believed the government would provide both a road out of the valley and a telephone line. In March 1895, Rev. Saugstad went to Victoria and returned with $10,000 for road and bridge work. With shovels, axes, crowbars, peevies, brute strength and determination, the men whacked out a wagon road up the valley and built a 600-foot pole bridge over the Nutsatsum River. They were paid a dollar a day for their efforts. In May, sixty more people arrived, mostly wives and children. Bella Coola was in her spring mode then, and the colonists realized what Jacobsen and Saugstad saw in the place. The wealth was there. Fish and game were plentiful, there was an abundance of wild fruit, the mountains backed off and let the cheery summer sun warm the valley and the colonists' hearts. They had the odd skirmish with bears, but they got used to the mountains watching over their shoulders. Land clearing was a slow, backbreaking business, but once they had the huge trees and underbrush off, the soil was rich. They built handsome log houses and planted gardens in the little patches they liberated from the forest.

In October 1895, Clayton took the first team and wagon up the new road. It was loaded with supplies to help the 220 colonists celebrate their first anniversary. Twenty miles of wagon road went up the valley. The main settlement, called Bella Coola, was in the middle of it, a two-day hike from tidewater. Hagen Christensen had a store and post office there, the church was finished and a school started, the latter a tent with a big stump in the middle that served as a desk for Iver Fougner, the teacher. Some colonists settled farther up the valley at Nutsatsum; others settled down the valley, about four miles from tidewater, an area called Lower Bella Coola. Later, to avoid confusion with the original settlement at tidewater, which was also known as Bella Coola, the Norwegians named the middle or main community Hagensborg, after Hagen Christensen. They didn't get around to changing the name of Lower Bella Coola, though, and to this day strangers are puzzled to find Lower Bella Coola four miles upriver from Bella Coola.

Rev. Saugstad died in March 1897 after a short illness. He left his young wife, his large family, and his dreams. Those dreams were never realized: Bella Coola did not become one of the province's major agricultural centres, run as a co-operative by morally correct people. But the colony took hold. The men found their future was in commercial fishing, and by the early 1900s Bella Coola had two fish canneries. The colonists had stores, hotels, a bank, a newspaper and a number of sawmills, and a community-owned telephone line ran thirty miles up the valley. (Customers bought their own phones and the switchboard was in the operator's kitchen.) A steamship from Vancouver called once a week, bringing supplies and settlers on their way to Bulkley Valley, and those settlers made regular trips back to Bella Coola to get supplies. The fur trade was brisk. Some settlers trapped, as did the Nuxalk, and people from the Anahim Lake area came down the mountain with their furs. One trader, Southern Carrier Antoine Capoose, made yearly trips from the west Chilcotin with as many as ninety pack horses laden with furs. John Clayton had his own steamboat to carry supplies between Bella Coola, Bella Bella and Victoria.

# CHAPTER 4

# All That's Missing Is a Road

*The great hindrance to rapid settlement and development of this vast but little known portion of BC is the almost total lack of transportation facilities, there being practically no waggon roads at present and only a few pack trails. There is a considerable extent of agricultural and pastoral land but a railway is considered necessary to open the country. In the meanwhile, a few enterprising pioneers have penetrated the fastness and are establishing homes for themselves and these speak enthusiastically of the great resources of the country and of its splendid destiny.*
—Report on Chilcoten, Bureau of Provincial Information, 1907

The lack of decent roads in Chilcotin was one problem for settlers, getting across the Fraser River was another. By the time the road to Hanceville was officially recognized in 1888, there were three ways across the river, none of them satisfactory. Soda Creek was the official gateway from the Cariboo Waggon Road, but this northern route was the long way around. Cattle drives went south through Gang Ranch and Clinton to the railhead at Ashcroft. The third route, the father of Highway 20, left the Cariboo Waggon Road at 150 Mile House, swung south of Williams Lake, crossed the Fraser at Chimney Creek, climbed to the plateau, and met the other roads at Fred Becher's place.

Gordon Farwell, a pioneer who settled in Chilcotin in the 1880s (Farwell Canyon and Farwell Creek are named for him), described his first Chimney Creek crossing as "quite a picnic." He travelled with Frank English who drove stage for Johnny McGinnis. McGinnis met English with a canoe and took freight and passengers to the Chilcotin side, where he loaded them into an express wagon for the journey to Becher's. "Frank had breakfast out of a bottle," Farwell wrote. "When we reached the long hill (three miles going down to the Fraser and when I say hill I mean hill, it's rocky with stumps and holes in the road too) Frank said 'Now I'll show you some wild west driving' and letting off yells and war whoops he put a whip to the team. Some of the road was cribbed up with logs and more than once the wheels on my side were running over the top of the cribbing no more than three inches from the edge. There was a drop of some fifty to one hundred feet. There were right-angled and hairpin turns which Frank, or rather his lucky star, manipulated at a gallop with much popping of the whip and bloodcurdling yells. When we reached the bank he continued to drive through the boulders to the water's edge, the Fraser being high right then.

"Johnny was waiting for us with the canoe and did he ever turn loose and cuss Frank for driving like that over the boulders," Farwell continued. "After an inspection we found two wheel spokes broken which didn't improve relations between the two but I thanked my lucky stars I was safely out of it with only two broken spokes and not have the wheels crumple up and send us flying over the grade onto the bank of the river. Frank made restitution by producing a bottle of scotch. A good brand cost $1.25 a bottle, everyone had one and gave you a drink."

Emedee Isnardy, who started his Chimney Creek ranch in the 1850s, operated a ferry for hire, but residents eventually persuaded the government to provide a 12 x 6-foot flat-bottomed scow ferry. Murdock Ross, a rancher who had fallen on tough times, was hired to run it for $500 a year. In winter the river froze in the canyon and the ice was usually thick enough to hold a team and sleigh. It was Ross's job to find a safe passage. When there was an open channel in the middle Ross helped nature along by throwing sticks and branches in it. If the ice wasn't safe, nobody crossed. The rest of the year travellers had to take their wagons apart, load them unto the ferry, and drag the horses behind. Nervous horses fought the tow rope and drowned. Crossing was especially exciting in high water when everything from animal corpses to uprooted trees came barrelling down the canyon. The trees would up-end, disappear, then pop up again like carousing porpoises. The acrobatic timbers must have been responsible for an item in the *New Denver Ledge*. "Crossing the Fraser River coming from the Chilcotin, Jack McGinnis, a stage driver, says he saw a fish sixty feet long which he thinks must have been a sea serpent. It must have been a floating tree for even the whiskey along the Cariboo Road is not strong enough to raise a fish of that length."

By 1897 the river road to Hanceville had outlived its usefulness. People settled north of Bald Mountain, out of its way, so Tom Hance built a new road from Sawmill Creek. He chopped ten miles straight through the jackpine forest before dropping over the bank to his place. The stretch is still called Hance's Timber although there isn't much timber left. From there the track wandered its way to Tatla Lake and was officially called the Ashcroft Trail. There were about seventy-five white settlers in Chilcotin at the time, and the provincial government recorded two hundred Tsilhqot'in farmers. Apparently non-farming Tsilhqot'ins didn't count. There were 16,000 cattle, many of them belonging to Gang Ranch, and 4000 horses, many of them running free. Ranches and homesteads dotted the eastern plateau and followed the river west. They were miles apart, many of them off the beaten track, and the population thinned as it went west.

Settlers groused about the Chimney Creek crossing, even though few of them used it more than once or twice a year, and in 1898 Fred Becher spearheaded a petition asking for a bridge. All forty eligible voters (male, white, British subject, twenty-one years or older) in Chilcotin, Williams Lake and 150 Mile House signed the petition, convincing the government a bridge would bring prosperity to Chilcotin and the province. By 1901 the preliminary studies were done and engineers had chosen a site ten miles below Chimney Creek. The firm of Waddell and Hendrick designed a unique bridge for the crossing. It featured a lower cable and counterweight to provide rigidity and overcome movement and "windlift" common to suspension bridges of the day. It was a handsome structure with a 325-foot centre span between graceful wooden towers, an 80-foot Howe Truss on the east side, and a 225-foot trestle approach on the west.

Work began in late summer 1902. Sam Smith was in charge of the project, which employed up to sixty workers. Little is known of Smith except that he did a remarkable job. It was impossible to bring in proper machinery, and almost impossible, to say nothing of costly, to get skilled workers to the site and keep them there. Materials were hauled from the railhead at Ashcroft by team and wagon or sleigh, and the road was horrible. Several hundred huge granite stones, weighing 350 to 450 pounds each, were brought in for the piers. The towers were fabricated from pre-cut and pre-numbered wooden timbers, and when they were unloaded at the site, Henry Durrell organized the pieces for the workers to assemble. A young ranch hand with more education than most, Durrell stayed in Chilcotin to establish his own small kingdom. Each of the 700-foot long, 2 ¾ inch suspension cables was shipped from Ashcroft in its entire length by a convoy of sleds or wagons, each carrying a loop. Manoeuvring the wagon or sled train with its cable umbilical cord must have been a chore. There is no record of who did it, but there is a record of what the cables did to Premier Edward G. Prior.

Prior was British gentry. A mining engineer and colonel in the militia, he was a strong supporter of the Canadian Northern Railway which was looking to link the interior with the coast via the Cariboo and Bute Inlet. He was elected to the legislature in 1887 and his political career included a federal cabinet post and, later, a stint as Lieutenant Governor. He was serving as Minister of Lands and Works as well as premier in 1903 when the Chimney Creek bridge was built. All materials for this bridge went through the government's standard tendering process with the lowest bidder getting the contract. Perceiving a conflict, the manager of the premier's company (E. G. Prior & Co.) didn't submit a bid for supplying the cables. Prior did—after he'd seen the others. Needless to say, his bid was lowest and his company got the contract. There were screams of rage from the legislature. Prior wouldn't resign, so Lieutenant Governor Sir Henri Joly de Lotbinière relieved him of his duties. There is no evidence Chilcotin cared one way or the other about the premier's fate, but the incident is of historical interest, not only because he was the only premier ever to be fired, but because his successor, Richard McBride, introduced the party system of government which changed the face of BC politics.

*Anna and the bridge. Anna Graham and the Sheep Creek Bridge are linked forever in the minds of Chilcotin old-timers. When Mrs. Graham came to Chilcotin in 1889, at age twenty, to marry her childhood sweetheart, she made the trip from Ireland alone. On the last lap she had to cross the Fraser River in a flat-bottomed ferry boat. A rambunctious ram rode across with her. It must have been quite a trip because Anna vowed never to get in a ferry again, and she didn't. She left Chilcotin for the first time in 1912, via the bridge. This photograph dates from the late 1920s when Anna (right), her husband Alex (background) and a friend were on their way to attend the graduation of their daughter Kathleen from nursing training.*

*Becher House. This wayside inn, store, post office and saloon was the nerve centre of Chilcotin for years. After it burned in 1915, Fred Becher (pronounced "Beecher") built a new one—very posh and very large. It is shown here after a renovation in the 1940s.*

*Off for a visit. Distances did not deter the Bayliffs–Hugh, Gertrude and son Gay–from visiting neighbours. Their closest neighbour, the Newtons, lived only a few miles away, but it was a major undertaking to get to Alexis Creek. Mrs. Bayliff is always pictured, as here, in elegant dress, even when she was out riding in the bush. Given the living conditions in Chilcotin at the turn of the century, keeping so well groomed must have been a challenge.*

*Chimney Creek Bridge. Petitions and complaints from local residents persuaded the provincial government to build a bridge across the Fraser River in 1903. It was a massive undertaking: there was no proper road to the site, all the building materials had to be hauled in from Ashcroft, and skilled workers had to be imported. The bridge, opened in 1904, was officially called the Chimney Creek Bridge. It was also called the Chilcotin Bridge and the Fraser River Bridge. When they weren't calling it something vulgar, Chilcotin residents called it Sheep Creek Bridge. (BCARS 73022)*

The bridge opened for traffic in September 1904, at a cost of $65,000, almost double the estimate. "It is fifteen years since this great agitation in favour of a bridge commenced and although it is true that hope deferred maketh the heart sick, it is equally true that everything comes to those who wait," the Ashcroft *Journal* commented on August 20, 1904, in announcing the opening. Over the years the span was called many names, most of them obscene. Government documents of the day called it the Chimney Creek Bridge, although it wasn't near Chimney Creek. It was also known as the Chilcotin Bridge and the Fraser River Bridge. Settlers called it Sheep Creek Bridge because the hill on the west side is called Sheep Creek Hill. One story says there was a sheep ranch on the hill, another says there were wild sheep there. Sheep Creek itself is also known as Sword Creek for Tom Sword, a harness maker who lived near it before the turn of the century. Mostly the bridge was called the Bridge. It dangled a hundred feet above the Fraser and, in spite of its unique design, it swayed. It swayed a lot. It swayed and squeaked and squawked and the noise and the movement spooked every living thing that crossed it. Cattle were wild-eyed and balky, horses were skittish, sheep refused to cross. Terrified travellers trembled across it and even the stout-hearted breathed a sigh of relief when they reached the other side. There were holding corrals for stock at each end of the bridge, and a gate on the west side. In 1915 Public Works posted a notice on the bridge ". . . vehicles not faster than a walk. NOT more than 25 head of stock to be allowed on the bridge at any one time. See they are strung out." It wasn't deemed necessary to limit the number of vehicles on the bridge at one time; the roadway was only twelve feet wide and no one was fool enough to drive out on it when it was occupied.

The new crossing eliminated the hill on the east side that alarmed Farwell, but it created a monster of even greater proportions on the west side, a narrow track that inched up the towering canyon hill to the plateau 1200 feet above the river. The Hill, as it was known, was four miles long, with a 20 percent grade (a 10 percent grade is the maximum now) and six switchbacks. The reverse turns were steepest at the corners and the worst one was just up from the Bridge. It hung out over nothing and was dubbed

Cape Horn. Like everything else in Chilcotin, the nature of the Hill depended on the weather. It was bad enough when it was dry; it was dreadful when rain or snow made it slick or when it was icy, which it was all winter. When it was slippery, westbound traffic had trouble getting up it—if you stopped, you stayed stopped—and eastbound traffic had trouble staying on it. Anything with tires wanted to shoot off the edge.

Once on top, travellers were faced with Becher's prairie, a huge, open, undulating, rock-strewn plain. Cariboo writer and politician Louis Le Bourdais likened it to a vast feeding ground, because at first glance, the thousands of greyish rocks looked like grazing sheep. One old-timer explained the rocks were carried by a glacier who "got tired and set a spell and when she got going again she just left them rocks there." The prairie was formidable in winter because the winds had a good sweep at it and whipped the snow into incredibly deep drifts.

The bridge did not put Chilcotin on the map nor develop its splendid destiny. It didn't lessen distance to the railhead or bring a market. It did give cattle buyers easier access to the western ranches, but it was still shorter to drive the animals to Ashcroft via Gang Ranch. A twin bridge was built at Churn Creek, the Gang Ranch crossing, in 1912.

The bridge didn't do anything to improve the road in general, either. Chilcotin settlers still had to battle the rough roads to get to it, and that was always a major challenge. It did make life easier for Tom Hance, who had the mail contract, and anyone wishing to do business at the 150 Mile House. They could now go directly, without making the loop around Soda Creek.

CHAPTER 5

# The First Link

*Within the area bounded by Alexis Creek thence to the west end of Tatla Lake, including the lands and valleys in the east and west branches and the main Homathko, Choelquoit Lakes and tributaries to Tatla and Chilko, there are 80,000 acres of good unoccupied land. This does not include cattle range or the number of places with houses built and abandoned after a year by settlers who could not abide the solitude and distances. It is expensive to go to Ashcroft, impossible to get through to the coast, and there is no near market. A trail to Bella Coola would help.*
—Surveyor A. L. Poudry, 1896

The depression of the late 1890s put many of the smaller ranchers in the Riske Creek area out of business. Their places were taken up by the larger and more affluent landowners such as R. C. Cotton and Charlie Moon. Cotton was an English aristocrat who mudpupped for Old Drummond in 1897, then bought his ranch (the original Riskie and Withrow places) and others as well. Cotton called his holdings, which included 1500 head of cattle, The Ranche. People called him the Chilcotin Cattle King. It took Moon, another Englishman, longer to acquire his empire, but he ended up with more. Moon went to Chilcotin in 1888, when he was sixteen. He worked on different ranches and in 1902 bought Deer Park (the original Doc English place). By the 1930s he had 3000 head of cattle and thousands of acres of land. By the turn of the century more enterprising pioneers had penetrated the fastness of Chilcotin and settled in the valley between Hanceville and Redstone. By 1907 there were 121 settlers on the plateau, a net gain of 46 people in a decade. Some of the newcomers were wives, some were children born to settlers, and some settlers may have been counted more than once. The *BC Directory* listed Benny Franklin three times in three dif-

ferent places—Hanceville, Alexis Creek and Tatla Lake. On the
other hand, some settlers lived on land a long time before they
got around to acquiring ownership, and like the non-farming
Tsilhqot'ins, they may not have been counted at all.

The directory did not count anyone living west of Tatla Lake.
Lower Chilcotin and the valley were heavily populated in com-
parison with the west end. In 1907 there were a handful of Native
people living in the Anahim Lake area—a Tsilhqot'in family, the
Sulins, who hadn't moved east after the war of 1864; and a few
Southern Carriers including Chief Domas Squinas and Antoine
Capoose. There were half a dozen white bachelors, and three
Norwegian colonists who left Bella Coola for the high lands of
west Chilcotin. Settlers found this vast area west of Tatla Lake in
the 1890s. None were cattle barons, the land wasn't up to it. They
mostly raised horses. Among them were Pat McClinchy, a mite
of a man, an old windjammer who sailed around Cape Horn.
Every time he had a drink he'd go on and on about it. He lived
by the Kleena Kleene River about twelve miles past Tatla Lake.
George Powers, an American, had over 2000 head of horses.
When he eloped with Jessie Bob, Chief Anaham's granddaughter,
her father sent thirty riders to get her back. The lovers fled to
Charlotte Lake, south of Towdystan, and the pursuers gave up.
Sam Colwell, Frank Render and Adolph Schilling went to
Chilcotin from the Omenica gold rush. Colwell staked a claim on
Perkins Peak, sold it for $18,000 and bought property by the
Kleena Kleene River. Render ranched at Kleena Kleene and
Schilling founded Three Circle, the first ranch in the Anahim
Lake area, thirty miles from Engebretson. George Turner, an
American of some mystery, showed up in the 1890s. Everyone
was convinced he was on the run from something. He carried a
pair of six-shooters tucked inside his shirt.

The Norwegian colonists who headed for the high land were
Tom Engebretson, who took up land at Nimpo Lake, and Jakob
Lunaas and his daughter Annie, who settled at Towdystan. Tom
and Annie later married. Their oldest son, Fred, would always
shake his head when anyone asked why his father and grandfather
left the warm, fertile Bella Coola valley and their fellow colonists
for the cold, barren, lonely west Chilcotin. "They must of wanted

lots of land and didn't care to clear those big cedar trees and all that brush in Bella Coola," he would say. "There was big talk of a railway through here from Bella Coola in them days, maybe they thought it would make them rich." Fred would ponder the question, shaking his head, his bright blue eyes wide with wonder. "There was no other reason to come here unless they were out of their minds," he'd conclude. "This is a ridiculous place to be. The only way you can make money from land around here is to sell it to some damn fool who doesn't know any better."

Fred had a point. West Chilcotin doesn't really call you. It is high country, around 4000 feet above sea level. The growing season is scrimpy, the winters endless, and the grass that grows under the scraggly jackpine trees is sparse and low in protein. The jackpines grew as thick as hairs on a bear's back and while they made good log cabins, fence rails and firewood, they had no market value. What west Chilcotin did have was plenty of elbow room. A man could be free. As Fred put it, he could take a leak wherever he pleased without offending anyone. He might even make a living from a swamp grass meadow if he didn't want too much. The Drummer's beat was loud and clear in west Chilcotin.

Towdystan, the Lunaases' place, was ten miles east of Fish Trap. Fish Trap has a special place in Chilcotin history. This bend of the Dean River was once a favourite Tsilhqot'in fishing spot. It was there, in 1864, that Alfred Waddington's packer Alex McDonald and his men, warned of an attack by the warring Tsilhqot'ins, dug a trench on a knoll overlooking the river and prepared to defend themselves. They spent two days holed up in the earthworks, but when nothing happened, they made a run for it to Bella Coola. They were ambushed a few miles down the trail. A plaque marks the spot, and the depression made by the earthworks is still clearly visible.

Jakob Lunaas called his ranch The Forks because two old trails met there, but nobody much used the southern trail, fifty long miles to Benny Franklin's place at Tatla Lake, so the name didn't stick. Fred thought Towdystan was a mispronunciation or a mis-understanding of a Native word because it means liquid, which, he said, doesn't make sense. Lunaas rebuilt an old trail from Fish Trap to the Precipice, the mountain of basaltic rock that impeded the Palmer Trail. This route, called the Lunaas Trail, became the popular way to and from Bella Coola. Lunaas kept his property in the valley and didn't move to Towdystan to stay until 1907, about the time Annie and Engebretson wed.

A third Norwegian family, Ole Nygaard and his wife, also left the colony for Chilcotin, but they settled west of Tatla Lake, nearer Redstone. They used the trail on the north of Tatla Lake and the Kleena Kleene River to get to Towdystan. Annie Lunaas used to visit Mrs. Nygaard occasionally. The trip took two days on horseback, so she camped out overnight on the way.

The two ends of what is now the Bella Coola/Chilcotin Road had nothing to do with each other. Bella Coola residents' only interest in Chilcotin was as a way out. The two ends of Chilcotin didn't mesh either. The west Chilcotin residents looked to Bella Coola for supplies. The Nygaards shopped at Norman Lee's because it was relatively close, but the English gentry didn't go out of their way to befriend the Norwegian "cabbage heads," so they weren't really a link between the two. Residents of the Chilcotin valley and Riske Creek area looked east to the 150 Mile and Ashcroft. They knew—and cared—as little about their westerly neighbours as those neighbours knew and cared about them. The only thing the two Chilcotins had in common was the name and the wide blue sky.

The first settler to look both ways was Alex Graham's younger brother Bob, another veteran of the Omenica gold rush. He bought the Tatla Lake place from Benny Franklin in 1902. Graham, who became one of west Chilcotin's most influential citizens, could see that a road to Bella Coola would open up the country.

Bella Coola was at its peak in the days before World War One. Its destiny as a major port seemed assured. The Pacific and Hudson's Bay Railway was negotiating for property in the valley and in 1914 J. M. Rolston began surveying a route for the line. The Norwegians were a homogeneous bunch, but they were always forming little groups—farmers' institutes, political clubs, business associations. The different groups didn't necessarily get along, neither did the colonists. The Liberals, led by Barney Brinildsen—a storekeeper who published the valley's newspaper,

*Becher House, Riske Creek, late 1930s. The Bill Grahams drop in on their way home from Williams Lake.*

*Bella Coola, 1920s. Left to right: Bill Nelson, Adolph Christensen, his eldest son Andy, unidentified, Alger Brinildsen and surveyor J. M. Rolston. This photograph shows the original townsite north of the river. Marit (Mrs. Adolph) Christensen's restaurant is in the background.*

*Becher House, 1920s. Riske Creek area residents stop for their mail and groceries at Becher House. Left to right: Mrs. Stowell, young Willie Johnson. In the car: Emogene Stowell, Bill Owen, Dick Stowell (driving) and Clara Meldrum.*

*A. C. Christensen Ltd., Bella Coola, late 1920s. Adolph's store in "new" Bella Coola was built after the flood that destroyed the original townsite and caused residents to move across the river to higher ground. At left is Charlie Taylor's barber shop and jewellery store. During the war a repeater station occupied this site; later the present Co-op store was built.*

the *Courier*—feuded with the Conservatives, led by Adolph Christensen, the rival storekeeper. The one thing they all agreed on was the need for a road out of the valley. They wrote letters, petitioned and complained, and some of their comments were prophetic. In 1911, Brinildsen said that at the rate it was going, it would take two generations to get a road to the interior. It did. Noting the road crept up the valley at a rate of one and a third miles a year in the last seven years, an Atnarko bachelor said it would be twenty years before they got a road. It was longer than that.

The first link among the three settlements came in February 1913, when a message went over a fragile telegraph line connecting Bella Coola to 150 Mile House. Construction of the 300-mile line began in 1911. They say the construction foreman, J. Thorne, never walked, he ran through the bush blazing trail with an axe. Both the construction of the line and the final product added a few more jobs to the settlers' repertoire as well as linking them to the outside world. The line, a single Number Eight galvanized wire, was mostly strung on trees. Local men were hired to help the linemen, and they worked long hours. The wasps and mosquitoes nearly drove them mad but the wages were good, 21 dollars for a seven-day week plus room (tents) and board. Tom Engebretson hauled wire. Bob Graham traded teams and wagons for a hay crew. During the construction, three women came across a cache of insulators in Bella Coola and, having no idea what they were, took them home and used them for flower pots. They were caught and charged with theft but the judge let them off with a lecture.

When the line was completed, telegraphers and linemen were needed to keep it working. Mosher Creek in the Bella Coola valley is named for one of the first telegraphers. Sam Colwell wasn't the first man at the key at Kleena Kleene but he was there the longest. Bella Coola telephones were attached to the system and Chilcotin settlers who lived close to the line got telephones too. There were phones at Engebretson's, Graham's, Lee's, Becher's, all down the line. The line was too weak to transmit calls directly from Bella Coola to the 150, so calls were relayed, and people could always call the telegraphers who would send a message out on the key. Domas Squinas, a prosperous trapper who lived at Anahim, was one of the relayers and he thought the system was hilarious. He laughed so hard he could hardly pass on the messages. He and his friend Antoine Capoose would get on the line and yell at each other. Down country, the gentry's Chinese workers, who seldom had a chance to see each other, would visit on the phone in the evening whenever they could. The telephones were the settler's version of the moccasin telegraph: each one was equipped with a receiver, or "howler," which broadcast every call into every home within reach.

One person, a Mrs. Hicklenton, asked the provincial government to follow the telephone line with a trail. A prospector, Mrs. Hicklenton claimed to be the first person to take a team and wagon from the Gang Ranch to Tatla Lake via Hanceville in 1889. Her husband minded their two daughters while she was off in the bush seeking their fortune. She walked to Bella Coola many times and in 1911 she was so upset with the windfalls on the Lunaas Trail she wrote to H. Clemens, the Member of Parliament, asking for $500 to clear it. "I'm not a woman who wants to take anything from the men but I feel capable to handle this little job. I think I am the best one to ask to open this trail," she wrote. She said there should be as good a trail to Chilcotin as there was to Ootsa Lake. "The telegraph line is costing $68,000 and it wouldn't cost much more to build a proper trail following the line," she said, adding the trail would do until a road was built. Nobody acted on her offer.

The telephone didn't solve the road problem. Premier McBride's government was concerned over the high cost of extending roads to isolated settlements, and was looking at a comprehensive road system for the province. Roads came under the Department of Lands and Works until 1908 when the Public Works Department (PWD) was formed under a separate minister. PWD recognized the Chilcotin Road as far west as Tatla Lake (officially, the Ashcroft Trail), and in the fall of 1914 an engineer named Joseph Fal was sent to scout a route to Bella Coola. Render, Colwell and McClinchy had chopped out twenty miles of road on the south side of the Kleena Kleene River from Graham's, but Fal said it was "very bad," especially along

McClinchy's meadow, where one side of the road was two feet higher than the other and the bottom was soft. "It will always be bad and a change will have to be made to take the road away from the river, so do it now instead of wasting money," Fal recommended. His advice was taken fifty years later. He said the trail to Schilling's was good going, except for the drop off the plateau to McClinchy Flats; the Lunaas Trail was fine except for windfalls; and the valley road could become the finest in the province. Then he sang the song Lieutenant Palmer had written fifty years before—the four to eight miles of road over the mountain would be "tough going." Nothing was done with Fal's report. The valley road and the Chilcotin road kept going their separate ways, lurching from one settler's yard to another.

"If you have a good team, a strong wagon and a light load you can, by careful driving, go six miles further now than you could six years ago," Brinildsen wrote in 1913. Actually there was more progress on the valley road than there was in Chilcotin. In 1912 a bridge was built over Kahyhlst Creek, where Mackenzie had entered the valley so many years before. A number of Seventh Day Adventists took up property in the area and called it Firvale. When the bridge burned a few years later, the replacement was called Burnt Bridge and Kahyhlst faded into history. There are several versions of the bridge burning. The best is that a remittance man was exploring the valley when night caught him. He camped in the middle of the bridge and lit a fire at each end of it to keep the bears and beasties away. It did, but the bridge burned.

The Nutsatsum River ate bridges. It could be counted on to flood regularly and it kept changing its bed. PWD crews were hard put to find a safe place for a bridge, and people were peeved when the 1914 move added two and half miles to the trip up-valley. "The road superintendent now has a force of men double tracking the road in the neighbourhood of Nootsatsum. The road being double tracked means the poor dear settlers are being double crossed," the *Courier* sniffed.

The combination of the Grand Trunk Pacific reaching Prince Rupert and the outbreak of World War One put an end to the Pacific and Hudson's Bay's railway plans and Bella Coola's rendezvous with destiny. In October 1917, a flood devastated the townsite. One of the casualties was Brinildsen's newspaper. It was a few years before residents got around to pressuring for a road again.

CHAPTER 6

# Bed, Board and Jawbone

*Money is the root of all evil. Please give us a few roots.*
—Sign in Norman Lee's store

The Hudson's Bay Company failed in Chilcotin, but independent fur traders did well enough. As the population increased, most became general merchants. It didn't take long for settlers to realize the Tsilhqot'ins had a good thing going with the trapping, and it became the mainstay of the Chilcotin economy. Furs gave many ranchers their start and saw them through rough spots, often bringing in more money than their cattle did. The ranches needed huge tracts of land, and unless settlers had money to buy land, they had to turn to other ventures to make a living. Most ranches had a sideline, small holdings often were the sideline. Ranchers were storekeepers, innkeepers, postmasters, policemen, justices of the peace, trappers, loggers, and at one time or another, everyone made a dollar or two doing road work.

Most merchants became cattlemen because of jawbone, or credit. There was so little hard cash in the country merchants had to use a credit, or barter, system. Ranchers, big and small, settled their accounts once in a while, usually in the fall when they sold their cattle. Or, they paid their debts with a cow or two, or perhaps a horse. Others, including the Tsilhqot'ins, paid however they could. If they didn't have furs or stock, they traded buckskin work, or labour for goods. The system worked well for the merchant/ranchers as it provided them with stock and captive cowboys. It was the Cariboo version of owing your soul to the company store.

Some ranchers went into storekeeping on purpose, some in self-defence because neighbours kept borrowing things. Few merchants said no to a post office because it attracted customers. Most places added room and board to their offerings as traffic increased, and because settlers who travelled all day to get their mail weren't about to return home the same day.

Fred Becher had the first stopping place. When he bought out George Dester in the late 1880s, the establishment included a store, saloon and wayside inn, although the partners were fur traders first and foremost. As road traffic increased, the saloon/inn became more important. Becher called his place Becher House and for years it was a gold mine. All the roads met there, and no matter where they came from, travellers were badly in need of rest and refreshment to restore their battered bodies and shattered nerves. Becher's motto was "We aim to please—for a consideration." Becher House was Chilcotin's nerve centre for over forty years; in its heyday it was the grande dame of all Cariboo stopping places. It sat in splendid solitude in the basin of Riske Creek, surrounded by rolling grasslands broken only by patches of willow thicket and clumps of cottonwood trees crouching in the hollows.

Becher was a typical British innkeeper: cordial, kind, but tough when he had to be. He was a big strapping man, dark-haired, dark-eyed, with the bearing of a Queen's guardsman. Some claimed he soaked up liquor like a sponge but never showed it. While he never started a scrap, he never lost one, either. Known as the Den of Iniquity, the saloon was the scene of magnificent sprees. Men rode great distances to get there. Becher stayed open all night when there was a crowd and he gave a bottle of whiskey and free breakfast to anyone who blew three months' wages at the bar. He brought whiskey in by the keg and sold it for a dollar a bottle, seventy-five cents for a refill. A single drink cost fifteen cents, two for a quarter. He didn't make change but no one stopped at two drinks anyway. After a weekend in the bar patrons left with fat heads and flat pocketbooks, every one of them convinced he'd had one hell of a good time.

The story is told of how Becher came home one day to find the saloon a madhouse, full of men and their horses. Dan Nordberg, the bartender, was an old man then, not up to coping with rowdy crowds, so Becher waded in to clear the house. Nordberg darted out from behind the bar and caught his arm. "No, no, wait a bit," he said. "The boys have just been paid, don't you see. By morning we'll have every cent off them. The mess on the floor can be cleaned."

Becher had the best-stocked store north of Vancouver. He raised cattle, horses and a large sheep herd; his rams were said to be the most ambitious in Chilcotin. Numerous settlers got their start working for him. He was a shrewd businessman who seldom came out on the short end of a jawbone. In a typical transaction he acquired Ross Gulch for a debt of $600 and sold it for $3000. But Becher was best known for another deal. When he wanted to marry Mrs. Cotton's companion, an English gentlewoman, he traded his Tsilhqot'in wife to a man who owed him money. At the time this was thought to be considerate of him, as most men in similar situations simply dumped their Native wives.

The Drummer doesn't always appreciate success. In 1915 Chilcotin smacked Becher a good one. During a cold spell in January his place burned down. He took the loss stoically, although he had no insurance and lost $70,000 in the store alone. He had to build a bigger hotel than he needed to keep his liquor licence, and even though he cut all the lumber on his own water-powered sawmill, he went heavily into debt building a palatial thirty-room establishment. It had a conservatory (the plants froze in winter), a grand dining room and a ballroom with twelve-foot ceilings. The furnishings were imported at great cost from Liberty's in London, England, and included fine brass beds for special guests. Customers didn't always appreciate the swank because there was no indoor plumbing and the bedrooms were cold as sin in the winter.

Chilcotin had three major social events—a three-day spring race meet, held every May 24 weekend on Becher's Prairie, and two balls, held spring and fall at Becher House. Everyone attended these affairs, regardless of weather, distances or road conditions. The gentry competed fiercely at the races with horses as aristocratic as themselves. Doc English brought horses from Ashcroft for the races. The dances were equally popular. Settlers thought nothing of riding saddle horse or driving buckboard fifty miles in sub-zero weather to attend. They often made better time than the motorists, who found lodging in the ditch at least once coming and going. The dances were described as snappy, scrappy affairs. They began at 8:00 p.m. and ended with the home waltz at 5:30 a.m., with time out for a sumptuous midnight feast. They were Chilcotin's version of the debutante ball as they gave unattached lads and lassies a chance to look each other over.

The next stop on the road west was Hance's, twenty-two miles— a good day's journey—from Becher's. Tom Hance started as a fur trader, but he was also a roadbuilder, postmaster, mail contractor, policeman, jailer, host of a stopping place and a rancher on the side. Hance's was the only place in the valley for so long it probably became a stopping place out of necessity. Cattle buyers, salesmen, government officials, road crews, wanderers and local people stayed there, and for years it was one of the regular stopping places for teamsters and truckers. The Hances were famous for their hospitality and generous meals. They grew a big garden, so they had fresh vegetables and small fruit which were always a treat. Travellers raved about Mrs. Hance's baked beans and brown bread, and Chilcotin bachelors liked to go there for a good feed. Hances also had a flower garden with many of the plants coming from Nellie's family's garden in Victoria. Nellie must have been generous with slips of those plants because descendants of her yellow rose and lilac are in gardens everywhere from Williams Lake to Tatla Lake.

The Hances had five children—Grover, Judd, Hattie, Percy and Rene. Judd died of pneumonia when he was a young man. Hattie married Frank Witte, who came from the US, and she was the only one to have a family. The "Hance boys," as the three Hance sons were called, preferred ranching to anything else, and after Tom died in 1910, the place was best known as the TH Ranch (named for the Hance brand). The original Hance house burned sometime around the turn of the century, when the Chinese cook stoked the stove too much and caused a chimney fire. The second house had four bedrooms downstairs for guests, and a parlour upstairs. Mrs. Hance married again, this time to a Chilco Ranch cowboy, Jim Ragan. She died in 1935, and two years later the second house burned when Ragan upset a coal oil lamp.

The next year, the boys turned the TH into a guest ranch which became one of the most successful operations in the Cariboo. Its main attraction was the Hance boys themselves. They were

*Chilcotin Valley. Pioneers and politicians who thought Chilcotin would rival the Fraser Valley as the breadbasket of British Columbia were fooled by the good growing conditions along the river in the Riske Creek area and in the valley between Hanceville and Alexis Creek. In the valley the sun reflects off the rimrock along the north side and there are usually at least three frost-free months. It is possible, most years, to produce corn, cucumbers, peas, even beans and tomatoes. But a gardener on the plateau or in west Chilcotin counts himself lucky if a turnip survives. It can freeze every month of the year in most places. (BCARS 3698)*

*Lee's Place, 1920s. Norman Lee earned his place in history when he lost all his cattle in a disastrous attempt to drive them to the Yukon to sell the beef to miners. He returned to Lee's Corner and started all over again. After his marriage and the adoption of his son Dan, he sold the Chilcotin property to C. G. Temple and moved to Victoria. Temple couldn't get along in the Chilcotin and Lee had to take the place back. Lee's Corner was, and still is, the gateway to Big Creek, the Whitewater country, Nemiah and the new Ts'ylos Provincial Park. The store is on the right in this photograph. (BCARS 99859)*

*In the parlour. Hugh and Gertrude Bayliff and Tommy Young (r.) in the Bayliff parlour. In spite of the isolation of their ranch, the Bayliffs never lowered their standards.*

charming, courtly, and entertaining hosts. All three were flashy dressers, and along with the traditional blue jeans, cowboy hats and pearly-snapped western shirts, they wore gauntlets, neckerchiefs or string ties. When they were riding (and sometimes when they were not) they added jangly spurs and hairy chaps. They usually wore their riding boots, topping them in winter with fancy coverings. Grover's best, or visiting, hat had a bullet hole in it. No one ever knew the real story but he got a lot of mileage out of it. Men liked the Hance boys, but women were absolutely gaga over them, especially Rene. The attraction didn't wane as they got older, either. All three boys were unfailingly good-mannered to everyone, but Percy, who never married, never, ever lost his good humour, no matter what the provocation. "It's just as easy to be polite as not," he said. "I've never seen where getting riled solved anything."

Norman Lee's place was next on the road. The establishment he bought from Dan Nordberg in 1892 wasn't much in the way of outbuildings, but the trade was reasonably brisk and Lee did as well as anyone else. He also raised cattle, and he earned a niche in BC history when he took 200 head to sell to Yukon miners in 1898 and lost them all. He returned to Chilcotin with a blanket, a dog and a dollar, and started all over again. Lee had a quiet sense of humour but he didn't like rough language. When a cowboy cussed he'd admonish him. "Now, now, dear chap, that won't help, the cattle can't understand what you're saying." In 1902 he returned from a visit to England with a bride, his cousin Agnes. They adopted a son, Dan, and in 1913 they sold the place to C. G. Temple and moved to Victoria. Temple couldn't hear the Drummer and Lees had to take the place back in 1919.

Lee's was just six miles from Hanceville and they didn't really have a stopping place, although they fed anyone who came by at mealtime. They were better known for the ranch and the store. The store was Mrs. Lee's domain. When he outgrew Nordberg's little cabin, Lee built another store across the road from the house. Someone painted "This Is Lee's B.C." on the side of the store to greet westbound traffic. Known as Gan Gan to everyone in the country in later years, Mrs. Lee held court in the store until she was in her eighties and blind. She often served tea there, and she expected travelling dignitaries such as school inspectors and road engineers to stop and keep her posted on current events.

Gan Gan was one of the few settlers to learn the Tsilhqot'in language. Storekeepers needed the patience of stone with Native shoppers because they bought one thing at a time and could take all day doing it. Gan Gan did her knitting between sales. She thought nothing of leaving a customer in charge while she went for lunch, and she was always good for jawbone. When her eyesight failed she had her customers weigh and measure their goods and make their own change. "Few people will cheat an eighty-three-year-old lady," she insisted. When Princess Margaret, the Queen's sister, visited Williams Lake, Gan Gan was invited to the doings as the special guest of honour. She attended the official functions, but turned down a luncheon date with the princess because she had made a date with an old friend and wouldn't break it.

Chilcotin didn't short for Lees. Norman's brother Edward Penrose (E. P.) came to visit in 1888, heard the Drummer, and returned in 1895 to ranch at Redstone. E. P. was eccentric, even for Chilcotin. He had a classical education and among other accomplishments he knew several languages. When anyone was impressed by all this he'd snort and say it didn't make his cows breed any faster than anyone else's. His sister Helen, who worked for thirty years in San Francisco, came to keep house for him in 1921. Gan Gan's brother, Thomas Campbell Lee, came to visit in 1897 and he stayed too, settling in Alexis Creek where he built a store.

Chilcotin stores had a lot in common besides jawbone. They had attics full of furs and they all carried the same eclectic stock. "Everything from sewing needles to wagons," according to one. Perry Davis Pain Killer was a big seller. It was taken for ailments ranging from female complaints to hangovers and if a shot of it didn't cure the problem it helped the patient forget about it for a while—the main ingredient was alcohol. The stores were dark, cluttered caves redolent with the smoky smell of buckskin. Coils of rope, cowboy hats, bolts of gingham, and overalls fought for space with slabs of bacon or salt pork, coffee beans in metal boxes, loose tea in foil-lined crates and dried prunes and beans in gunny sacks. Kerosene (coal oil) came in four-gallon tins, two to a wooden box. Settlers coveted the crates—indeed, any wooden boxes. They used them for cupboards and shelves. They used the kerosene for medicine as well as lamp fuel, dosing themselves with it when they had the flu. Some swore it prevented gallstones. Yeast (baking) powder, big kettles, embroidery thread and beads were popular with the Tsilhqot'ins. The post office, if there was one, was a cubbyhole in a corner. Freight came from Ashcroft by team and wagon or sleigh until the railway reached Williams Lake.

Somewhere along the line, Benny Franklin acquired property between Hanceville and Alexis Creek, and with a partner, started a ranch and store. In 1908 he sold it to the Perry Martins who moved up from the US. Mrs. Martin kept the store for a time. Franklin moved on, but about this time his wife and their oldest son returned to the States. The youngest son, Major, later ranched at Nazko, and Benny built a road there, too. Alex and Anna Graham worked hard and prospered. After they bought the Alexis Creek property from Alex's brother Bob, they built an eleven-room house right beside the road, in what became the "downtown" area. The windows and doors were freighted in from Ashcroft; everything else was handmade. Anna was postmistress for over twenty years; Alex was justice of the peace until his death in 1934. Anna took paying guests (probably the mail customers) until Tom Lee married and added a stopping house to his establishment, which was just a hop and a jump up the road from Graham's.

For many years, the last stopping place along the road was Bliss's. It wasn't an official wayside inn, but local people often stopped for a meal or stayed overnight. William Bliss was a Boer War veteran who worked as a horse trainer on the Newton estate in England. A Newton son, Reginald, mudpupped for Hugh Bayliff, then bought a neighbouring ranch. When he decided to raise polo ponies, he had Bliss bring over an Arabian stud and two mares. Bliss took a liking to Chilcotin and he never left. He sent back to England for his wife and children, and took up property just east of Bayliff's. Local travellers liked to stop at Bliss's for a visit or a meal, sometimes overnight.

Stuart's store at Redstone was next on the road, a few miles up from Bayliff's Chilancoh Ranch. (Bayliffs did not cater to the travelling public). While Alexis Creek has about the best Chilcotin has to offer in terms of climate and growing season, Redstone is about the worst. It records some of the coldest temperatures in the province. (Although he spent most of his adult life at Redstone himself, E. P. Lee didn't think much of it. "Why anyone chose one of the coldest localities in Canada when there was plenty of opportunity in Kamloops, I do not know," he was heard to mutter on occasion.) There was a trading post at Redstone in the late 1890s, and Pete Stuart bought it in 1915. He died in a car accident in Vancouver three years later, and his brother Andy, a printer by trade and just out of the army, went to Redstone to settle the estate. He took so long doing it his wife Hettie, an English war bride, went to investigate, and she stayed too.

Andy was a bouncy little man with a bushy moustache. He "learned his furs" in short order and kept the business going. The Stuarts were hospitable but they didn't get involved with bed or board. Stuart's was the last store on the road until the late 1920s, and Andy often went west to buy furs or to deliver supplies. Sometimes he got sidetracked. On one trip to Sam Colwell's at Kleena Kleene the supplies included some jugs of home brew. It must have been powerful stuff: Andy was gone for three weeks.

CHAPTER 7

# The Automobile

*The first car of the season was driven to Hanceville, with some*
*difficulty, by a Kelly Douglas salesman on April 25.*
*—Cariboo Observer, 1920*

The year 1912 saw many changes in Chilcotin. That
year Anna Graham made her first trip out, crossing the Fraser
River on the new bridge. She'd meant it when she said she
wouldn't get in a ferry again. Her brother-in-law, Bob Graham,
married in 1912. He met Margaret, a young Scots widow, when
she visited her brother Sandy Robertson in Tatlayoko, just south
of the Graham place at Tatla Lake. Graham had intended to
propose to her, but hadn't gotten around to it when he heard
she'd left for Ashcroft to catch the train to go home. He leaped
on his horse and caught her before she got away. The Nygaards
sold the Chilanko Forks Ranch to Arthur Knoll, sight unseen, in
1912. Knoll, a German immigrant, farmed in Alberta before
finding his way to Bella Coola. Mrs. Nygaard carried her baby
down to Bella Coola in a basket on her back, and Mrs. Knoll
carried her baby up the hill to her new home in the same basket.

Another significant event was the building of the Alexis Creek
Hospital. Most place names, like Riske Creek, covered vast areas,
but a village developed at Alexis Creek early on. In 1912 valley
residents combined forces and built a log hospital between Alex
Graham's and T. C. Lee's. Dr. William Wright, a former naval
doctor, and his nurse/companion Miss Mary Good, staffed it. He
was the only doctor west of the 150 Mile House, and he made
house calls, travelling by buggy or sleigh. His trips of mercy
sometimes took several days. The doctor liked a nip and he
carried a bottle of refreshment under his coat, sipping the con-
tents through a surgical tube as he travelled. If the trip was a long
one, he could be snockered by the time he reached the patient,

and he often fell out of the sleigh or wagon on arrival. There is
no record of his ever losing a patient. One December the men at
Alexis Creek ordered a case of Christmas cheer for themselves
and had it delivered to the doctor so their wives wouldn't know
about it. They were hopping mad when they went to get their
bottles and found he'd "drunk every drop."

World War One cleared Chilcotin of young English bachelors,
who left to serve their country and didn't come back. The war
didn't stop progress, though. Some bureaucrat with a bizarre
sense of humour declared the Chilcotin Road officially open for
automobile traffic in February 1915. Fred Becher had the first
motor in Chilcotin, a 1915 Cadillac. He hired I. J. Purkeypile to
run taxi with it between the 150 Mile and Hanceville. Mrs. Cotton
of The Ranche was the first woman driver west of the Fraser
River; she had her own car. Alex Graham bought a Ford but his
family and friends weren't impressed. The vehicle looked like a
perambulator and his younger daughter Kathleen said it should
have had a push handle because she and her sister Frances spent
more time pushing, pulling and prying than they did riding. "We
called it the Fractious Flivver," Kathleen said, "and I'm sure it
kicked up its heels so much because it knew he was a greenhorn
driver." When Graham took Mrs. Macauley and Mrs. Bob Graham for Sunday drive, they got such a jouncing Mrs. Macauley
came back with a smashed hat and Mrs. Graham with a black eye.
The road certainly was not ready for motor vehicles.

The first cars arrived via the Gang Ranch Road, which wasn't
much, but the Chilcotin Road was worse. One bad spot went
through a meadow belonging to Anaham Reserve. The
Tsilhqot'in people adapted to the white man's ways as best they
could, and the Anaham Band was particularly progressive.
Thanks to Chief Anaham they had good farmland, and their
reserve was located on the main road which gave Band members
easy access to the white community. By the early 1900s Anaham
had its own sawmill. Reserve residents planted huge gardens and
raised cattle, horses and pigs for sale. They fed the pigs salmon,
so the bacon tasted fishy, but people got used to it. Men from
Anaham packed freight for the settlers, guided, and worked on
ranches, and they got along well as long as they "knew their

place"—few settlers considered the Tsilhqot'ins their social equals.

The residential part of the reserve was on a bench overlooking the meadow, and the road must have been a sore point, because in the summer of 1915, without any warning, Band members fenced it, obliging traffic to take a lumpy track along the bottom of the bench and through a ravine. C. G. Temple, who had Norman Lee's place at the time, discovered the fence and he was furious. He fired off a letter to J. Moffat, Cariboo district engineer, in Quesnel, demanding the "Almighty Natives" remove the fence. Moffat sent a hot letter to Indian Agent Isaac Ogden at Lac La Hache, demanding the fence come down. Ogden replied the Band couldn't cultivate the meadow with the road running through it. Moffat huffed back. He said the Band had eight hundred acres of "beautiful meadow land," didn't need the road, and didn't have the right to interfere with the King's Highway. He had the fence removed, spoke personally to Chief Anaham Bob, and thought the matter was settled. The next spring everyone in the village turned out, put the fence back, ploughed up a mile and a half of road and planted a crop in the meadow.

Moffat was flabbergasted. Letters flew. Nothing happened. The Band harvested the crop and planted another. The row went on for three years. When Moffat realized he wasn't getting the meadow road back, he demanded the Band pay $3000 to have the diversion brought up to standard. Ogden refused. Settlers grew tired of scrambling around the hillside, and egged on by Temple, they started signing petitions. (The person who used the road the most, Tom Hodgson, favoured the detour. He said the meadow road was mirey.) The row ended abruptly when someone in the Indian Department realized the meadow road had never been registered. Legally, it didn't exist. Moffat pulled in his horns in a hurry and found money to improve the diversion before the Band decided to fence it too. But Temple didn't give up. Sometime later he drove into the Anaham irrigation ditch and bashed his car. He wrote to Premier Duff Pattullo claiming damages. PWD bureaucrats duly investigated and found the ditch Temple had blundered into was beside the road, not across it. "Mr. Temple has a new car and doesn't know how to drive it," someone wrote on his file. Temple was a classic example of someone who didn't hear the Drummer. He went broke, which wasn't surprising given his attitude to the people from Anaham. They should have been his main customers.

Automobiles caught on in Chilcotin in spite of the road, which still ended at Tatla Lake. They didn't always have an easy time of it. The Cariboo *Observer* recorded one mishap. It seems Arthur Knoll failed to negotiate a turn on the Chilcotin Road, and his car landed in a gulch. Colonel Wycotte was camped nearby with a herd of Chilco Ranch cattle and went to the rescue. "A team was hitched to the rear axle and a cowboy roped the front axle and bumper. The big car was hauled out without much difficulty and the only damage was a punctured gas tank."

Traffic grew by bumps and bounds as fur and cattle buyers, government officials and salesmen joined local traffic on the road. In 1924, a young Englishman named Alex Marshall, who was dairy farming near Alexis Creek, built a garage between Graham's house and T. C. Lee's place, across the road from the hospital. Most ranchers were handy and repaired their own equipment, but motor vehicles needed gasoline and batteries, and by the time a car got to Alexis Creek, it was bound to have something broken, bent or shaken loose. It took the better part of a day to get from Williams Lake to Becher House on a bare road, another full day to Alexis Creek. Marshall had a freight truck too, and he built a stopping place beside the hospital.

That same year, Evan Jones was posted as district engineer at the 150 Mile House, beginning what was to be a forty-year affair with the Chilcotin Road. A gentleman of the old school, Jones was very proper, seldom surprised, and his sense of humour saw him through the rough spots. One of the first things he did was to appoint William Bliss of Redstone as Chilcotin's first permanent road foreman. Bliss was an experienced project foreman, he'd been in charge of improving the Anaham diversion, but now he was to keep the road healthy between spring and freeze-up. He was to grade the road with a multiblade drag (that took most of the summer, with time out for haying); clean ditches and culverts; brush and grub (dig out roots and rocks with a mattock); "turn water" (divert run-off water so it didn't wash the road away) in the spring; rebuild and repair roads as directed; and hire and

*Hanceville Bridge, 1928. Foreman Bill Bliss(l.) was in charge of this massive project. Among the workers were Cyclone Smith (in cowboy hat) and Riske Creek foreman Dave Chesney. Smith later lost his life at the 1932 Williams Lake Stampede, the only death in the event's long history.*

*Bull Canyon. There are a number of stories about Battle Bluff. They all mention an incident when some Tsilhqot'ins cornered a party of enemies on top of it and forced them over the edge. Battle Bluff overlooks Bull Canyon, so called because the Bayliffs and other early ranchers fenced the narrow end of the canyon to pen in the bulls. At one time the community held sports days, May Day picnics and jackpot stampedes on the flats in the canyon.*

*The Squinas Place, Anahim Lake. Although Chief Domas Squinas had a house at Ulkatcho, he spent most of his time at Anahim Lake where he and his family were the lone residents for many years. The Squinas place, shown here in the 1920s, is on a hill overlooking the Ulkatcho band's reserve beside the Dean River. The non-Native settlement grew up adjacent to the reserve. Most Ulkatcho people didn't have much to do with Anahim Lake until the Department of Indian Affairs built a school there in the early 1940s. Old Squinas, as he was known in later years, was a successful trapper and he raised cattle and a few sheep. But he was best known for his horses, the finest and fastest in west Chilcotin. Fur buyers and other travellers stayed at his place. Squinas, who spoke the Bella Coola language well, also had a camp at Stuie where he fished and picked berries in the summer. He died of injuries sustained when he was attacked by a moose he had wounded. (BCARS 94824)*

supervise crews. Settlers fought for jobs because it put cash in their pockets as well as improving their access.

One of Bliss's first jobs was rebuilding the Hance's Timber Road. For some reason Tom Hance was never paid for building it in 1897. It's unlikely he volunteered his time, as it was a major diversion. After his death in 1910, his widow set about correcting the oversight. She got nowhere going through regular channels, so in 1914 she consulted Cariboo MLA John Fraser. Public Works Department bureaucrats told him there was no record of the work being done and the account was of too long standing to settle anyway. Mrs. Hance left it at that until September 1923, when she received a bill from the land department for $270 plus interest owing on a pre-emption her husband had taken up in 1879. She immediately wrote to PWD pointing out "if one account was not of too long standing, the other should not be." PWD didn't agree and wouldn't pay. Rene, the youngest Hance son, entered the debate at this point but as he frequently feuded with PWD, the bureaucrats pigeonholed his letters. Mrs. Hance consulted Evan Jones in 1924. After investigating, he recommended she be paid, but Victoria held fast. So did Mrs. Hance. She wouldn't pay the pre-emption until someone paid for the road work. She didn't accept the argument about missing records. "The road is there, it has been used for thirty years, Tom Hance built it and he was never paid," she insisted.

Over the next few years four MLAs and numerous pioneers backed her claim. In 1928 Premier John Oliver, the attorney general, and the finance minister were involved, as well as public works and the lands department. Oliver wanted to pay but PWD wouldn't. When PWD caved in, the attorney general vetoed it. In 1929, Williams Lake's foremost citizen and merchant, Roderick Mackenzie, became MLA and he gave the system such a kick it dislodged the money. An order-in-council was passed to pay the Hance estate $1000 for the road work. However, the $270 owing on the pre-emption, $538 interest, and a $10 Crown grant fee were deducted. Eighteen years after she raised the issue and thirty-two years after the work was done, Mrs. Hance received $182 for the building of Hance's Timber Road.

Like the settlers, motor vehicles inched their way west, in most instances shuddering their way along the rough track long before it was appropriate. Few reached Tatla Lake before 1920. For the most part, settlers in west Chilcotin simply ignored the advent of the automobile. There were cars early on in Bella Coola. They were shipped in on the steamboat for those who could afford them and had the courage and fortitude to battle with the valley roads, which were just as bad as those in Chilcotin. Like the road west, the Bella Coola road petered out up the valley in direct proportion to the thinning of the population. Few people were rash enough to try motoring past Hagensborg.

CHAPTER 8

# The Longest Mail Run in the British Empire

*CHILCOTIN MAIL AND EXPRESS. T.J. HODGSON PROP.*
*Stage and Freight Trucks*
*Riske Creek     Hanceville     Alexis Creek*
*Whitewater     Redstone     Tatla Lake*
*Studebaker cars     Cadillac trucks     Reo Speed Wagons*
*Freight teams     Pack Horses*
—Advertisement, Williams Lake Stampede Brochure, 1927

He was said to be the best-looking driver on the Cariboo Road's Inland Express stage lines, but Chilcotin people remember Tommy Hodgson as a big man with a big heart. A Yorkshireman by birth, Hodgson came to BC's Fraser Valley in 1897. After working here and there he went to work at Cherry Creek Ranch between Cache Creek and Kamloops. One of his jobs was breaking horses for Inland Express, the company that hauled freight and passengers along the Cariboo Road. Before long he was driving for Inland, and in 1912, when he was twenty-seven years old, he moved to 150 Mile House and took the weekly mail run to Chilcotin. At the outbreak of the war, after he was turned down by the army, Hodgson took over the mail contract himself. Except for one year, he held it until he died in 1945.

When Hodgson took over the Chilcotin mail run the road was, by all accounts, in a deplorable state. Travellers said it was impossible to go more than seven miles an hour because of the rocks, broken culverts, doubtful bridges and huge chuck holes. W. W. Bell, the public works engineer, agreed. He didn't like the flimsy culverts ranchers used to cover their "very deep" irrigation ditches. He wanted to blast rocks out of the road, build proper culverts, bypass the sand banks between Newton's and Bayliff's, and bridge a deep gulch west of Borland Hill. None of it was done because the road from Isnardy's to the Chimney Creek bridge was in Cariboo District, from the bridge to Norman Lee's in Lillooet District, and from Lee's to Bayliff's in Cariboo again. It was difficult to sort out and nobody tried.

In the summer of 1915 a government official spent fifteen days getting from the 150 Mile to Anahim Lake and back. He rode saddle horse from Redstone. Hodgson went to Redstone every single week, carrying mail, passengers and freight. He used motor vehicles right from the start in the dry seasons. He started out with a Thomas Flyer, then in 1915 he bought a year-old seven-passenger Cadillac. He used a team and sleigh in winter for over twenty years because the Chilcotin road wasn't ploughed until the mid-1930s.

The 150 Mile was headquarters for "T.J. Hodgson, Chilcotin Mail & Stage Line" until Williams Lake came into being. In September 1919, the first Pacific Great Eastern Railway train puffed into Williams Lake, transforming what had been farmland for over fifty years into instant town. Within a year Williams Lake was bustling with banks, hotels, stores, and what was to become one of the largest stockyards and cattle-shipping points in British Columbia. The railway put an end to the brutal two-week cattle drives to Ashcroft, and Sheep Creek Bridge came into its own. Hodgson moved his headquarters to Williams Lake in 1920. He also moved his family. He had married Edythe Paxton, the daughter of 150 Mile House pioneers, in 1915. For years their home in Williams Lake was the gathering place for Chilcotin people who visited, partied and left messages there. They had two phones, one for themselves and one for Chilcotin. They kept a special bedroom for Chilcotin people; it was known as Mrs. Cotton's bedroom or Mrs. Graham's bedroom, or whoever was the current occupant. Mrs. Hodgson was a warm, gracious hostess; if she ever grew tired of all the company she never let on. The couple had six children—Jack, Wilfred, Marge, Phyllis, Patrick and Betty Jean—and their comings and goings kept things lively too.

The Hodgson establishment included a warehouse for the freight and a large two-storey building which housed the horses below and the trucks above. Hodgson spent as much time on the road as he did at home, but he hired drivers as well. Many a man sat behind the wheel of a Hodgson truck over the years. He did all his own monkey wrenching. One of his projects was to convert the Cadillac car into a truck. He cut off the back of it behind the seat, and bolted on a truck rear end. He drilled all the bolt holes by hand. The Chilcotin Road was barely eight feet wide in places, so truck boxes had to be narrower so they didn't hang up on trees.

The first obstacle Hodgson faced on his trips west was Borland Hill, going out of the Williams Lake valley. It was treacherous when it was wet or icy and both horses and vehicles grunted going up and over it. The road wriggled along to the Fraser River, crossed the creaky bridge, then shinnied up Sheep Creek Hill. It dodged the larger rocks on Becher's Prairie; they were rooted in hell and settlers had no way to move them. The smaller rocks were spaced to make the track as bumpy as possible, wheels went up and down on them like pistons. The road through Hance's timber was two ruts running between the jackpines, pitted with potholes and booby-trapped with what Hodgson referred to as "mirey" spots where trucks frequently bellied down. The hill down to Hance's was steep and sideling, fortunately it dried quickly. The road went along the bottom of the bench to Norman Lee's, then followed the Chilcotin River, winding around hills and gulches, through Anaham Reserve, Alexis Creek, Bliss's, Bayliff's and Redstone. Redstone was the end of the line until the late 1920s when Hodgson got the mail contract to Tatla Lake and Kleena Kleene. The 340-mile round trip from Williams Lake to Kleena Kleene gave him the longest mail run in the British Empire.

Kleena Kleene was Hodgson's last regular stop until the late 1940s. With the extended route, he travelled from Alexis Creek to Chilanko Forks one day, then to Mackill's One Eye Lake Lodge at Kleena Kleene. He stopped at Tatla Lake along the way. He usually stayed at Bob Pyper's, but he bought property near Chilanko and set up a tent for passengers. He stayed in it himself when Pyper was cranky. The tent sat on a wooden frame and was quite comfortable.

In the early days Hodgson tried to get to Becher's for his first

stop but sometimes he only reached Mackay's. Donald Mackay
had a ranch and stopping place at Four Mile Creek, east of the
Fraser. He charged fifty cents for a meal, fifty cents for a bed.
When cattle drives stopped over, ranchers often paid with a cow.
Mackay, one of the young settlers who got their start thanks to a
loan from Alex Graham, spent his later years working for Public
Works as a district engineering assistant.

Hodgson kept spare horses at Becher's. He used two light
teams to haul mail, six heavy teams for freight. In 1929, according
to the posted schedule, every other week Hodgson and the mail
left Williams Lake at 8:30 a.m. Tuesday, arriving in Riske Creek
at 12:30, Hanceville at 4:30, Alexis Creek at 6:00. Hodgson (or
his driver) stayed overnight at T. C. Lee's and left Wednesday
morning for Redstone at 7:00 a.m.. He arrived there at 9:00, Tatla
Lake at 4:00, and Kleena Kleene at 6:00 p.m. The process was
reversed Thursday, the truck getting back to Williams Lake Friday
evening. The next week, the truck left Williams Lake at 8:00 a.m.
Wednesday for a round trip to Redstone, and got back Friday
afternoon. The trucks always left Williams Lake on schedule, that
was the only guarantee.

Hodgson was pleased, but surprised, when he made a trouble-
free trip. The road was just earth, not gravel, so it reacted with
the weather. If it didn't rain in spring or fall, it was passable. It
was best at freeze-up. During breakup the trucks travelled on the
frost but they had to contend with frost boils, mud holes and
washouts on into spring. Many times a driver had to unload the
freight to lighten the load. Hodgson once loaded and unloaded
the truck twelve times in twelve miles of mud. He carried an axe
and a crosscut saw and cut down acres of trees to corduroy acres
of swamp. The wind frequently dropped trees across the road,
especially in spring, and once Hodgson had to beaver his way
through a mile of tangled windfalls. "We make it by the grace of
God—and horse manure to give us traction," he was known to
say.

If the road behaved, his vehicles didn't. They took a pounding.
As well as drums of extra fuel, he carried a good supply of spare
parts. In winter, snow filled the rough spots in the road but metal
crystallized and snapped and there were no such things as anti-

freeze or block heaters. In freezing weather, Hodgson drained
the radiator each night. In the morning he'd heat water (on the
stove in Tom Lee's store if he was at Alexis Creek) and pour it
over the block. When the block was warmed up, he filled the
radiator with hot water and start cranking. If it was really cold
he'd drain the oil and heat that too, and make a little fire under
the truck with fuel-soaked rags or whatever he had, to warm it
up. Freezables had to be unloaded each night, and bedding the
truck down at night and getting it going again in the morning
were major chores. Trucks often travelled with nose bags over
the radiator. The truck boxes were open on top and at the end,
and once the freight was loaded, a tarpaulin was tied over the
top. The canvas and the ropes froze stiff as did the driver's fingers
as he wrestled with them and the freight.

Becher's Prairie was a horror in winter. Chilcotin settlers stayed
away from it if they could, but Hodgson tussled with it every week.
The ten miles between Moon's ranch at the top of the hill and
Becher House was the worst stretch. The wind had plenty of
room to get up speed and it really swiped the snow around. Drifts
piled up all over the place. When the road drifted in it was every
man for himself. Motorists simply headed off into the prairie to
make a new track, hoping they wouldn't be ambushed by rocks
pretending to be snowdrifts. Sometimes there was no way around
the drifts, which meant shovelling through them. It took a lot of
shovelling. On one trip, Hodgson and one of his teamsters spent
a week fighting their way across the flats. Each had a loaded
freight sleigh pulled by a four-horse team. The animals managed
to lug the sleighs up Sheep Creek Hill, but the prairie was smoth-
ered in wind-packed snow so deep they floundered and fell in it.
The men tried every which way to get going. They thought they
had it right when they hitched all eight horses to one sleigh, but
the snow was sugary under the crust and the sleigh sunk in it.
They kept unloading freight until they got the right combination,
then they took one sleigh for a little way, unloaded it and went
back for another load, repeating the process until they had all the
freight across. Each night they'd take the horses back to Charlie
Moon's place on the Hill. One Hodgson driver, Frank Snyder,
died of exposure crossing the strip in the winter of 1918.

*Tommy Hodgson. He was the first trucker on the Chilcotin Road. Hodgson
trucks were Chilcotin's lifeline for almost fifty years.*

*Hodgson's place, Williams Lake. When Public Works was building the bridge at Hanceville in 1928, the department hired Hodgson to deliver twelve bridge timbers. Each 72-foot timber weighed close to 6000 pounds. To get them around the hairpin corners on Sheep Creek Hill, Tommy mounted them on two trucks, the stripped-down Cadillac and a Day Elder. The site of the Hodgson home is now a shopping mall–Hodgson Place. (BCARS 94628)*

*The Cadillac, 1920s. Tom Hodgson was licensed for passengers, express and the mail only–no freight. His first motor vehicle was a Thomas Flyer, but it wasn't successful so he bought a 1914 Cadillac touring car, pictured here. Later, once he was licensed for freight, Tom added a truck box and the Caddy ended its days hauling along with the rest of the fleet.*

Hodgson's trucks—he eventually had three—were Chilcotin's lifeline. Settlers planned their lives around their arrival. No one was upset if they were late, they always got there. People heard the older trucks coming six and seven miles away on a clear day, the gears growling, the motors rumbling, the solid rubber tires clacking on the stony ground. As well as making official stops at post offices and stores, Hodgson made many unofficial stops, delivering freight to settlers along the way. He'd wait at each post office until the mail was sorted, then take the individual sacks of mail and hang them on the fence or wherever it was convenient for settlers to pick them up. On the way back he'd pick up any outgoing mail. He didn't charge for this service and it added time to his trips. Often settlers would phone in a grocery order and the storekeeper/postmaster would stick whatever it was they needed in the mail sacks too. People sent to town for this and that and Hodgson or his driver went all over the place picking things up. Nothing was too much trouble, and Hodgson expected his drivers to give the same service. The trucks hauled groceries, gasoline, machinery, people, cases and bottles of liquor, anything anyone wanted or needed. This diversity gave government officials a headache because there was no licensing category for him.

Hodgson carried some awesome loads. In 1926 he took twelve bridge timbers to Hanceville. Each of the timbers was eighty feet long and weighed several tons. He used two trucks, one of them the stripped-down Cadillac. He built high bunks on each truck and carried one end of each timber on each truck. The Cadillac had been an open touring car, so there was no cab or roof over the front seat. Hodgson drove under the timbers in it, and Claude Huston drove the lead truck. It worked. They navigated the Sheep Creek Hill hairpin curves without mishap.

On one occasion, Hodgson built a double deck on the back of the Caddy truck so he could haul bigger loads with it. He was bringing sheep in from Chilcotin when he had a misadventure with it. Someone closed the gate by the bridge at the bottom of Sheep Creek Hill, and when he came around the corner he had to run up on the bank to miss it. The Caddy stopped but the top layer of sheep didn't. They landed in the front seat with Hodgson and his passenger. There was much swearing and some loss of dignity, but no injuries.

Hodgson was a tall, hearty, clean-shaven man. He enjoyed a party and thought nothing of driving his wife and friends from Williams Lake to the Becher House dances. He always wore high-laced boots and his shirts usually had a black line across the front from the steering wheel. Mrs. Hodgson would say, "the steering wheel won another round." He was generally good-natured but when he lost his temper he tended to throw his tools around. Once when he'd finished digging himself out of a snow-bank he chucked his shovel and it went over the bank. When he went to rescue it, he tumbled down the slope. He was wearing a buffalo hide coat and the snow stuck to it. By the time he'd clambered back on to the road he was one big, angry snowman.

At Alexis Creek he stayed at Tom Lee's. Mrs. Lee, a stickler for manners, set an elegant table with silver and fine china, and she wouldn't let men in her dining room without a jacket. This irked Tommy. He threatened to go to the table wearing a jacket but no pants, but he never did.

Hodgson was on the road until he was sixty, when he was felled by a series of strokes. His boys carried on the Hodgson tradition. They ran the business, providing the same kind of personal service, for another seventeen years. Trucks with the name "Hodgson" on their sides still haul into Chilcotin, although no one in the family is involved any more.

# CHAPTER 9

# The Country's Getting Crowded

*The first batch of men to settle in the Anahim area were explorers. The men wanted freedom to move around, they wanted to see what was on the other side of the mountains. It took some of them thirty years to settle down.*
—Mickey (Mrs. Lester) Dorsey

Settlers were well established in lower Chilcotin by the 1920s. They were part of the world. They had the railway at their doorstep, weekly mail delivery, motor vehicles, road foremen and schools. There were dances, polo games and cricket for entertainment, and Fred Becher had a radio. But, like the road, civilization took its time going west.

In response to nagging from Bella Coola, the Public Works Department sent an engineer named J. Carruthers to look for a road to tidewater in 1920. His report said the grazing commission should study the area west of Pat McClinchy's to ensure the land was suitable for stock raising before a road was even considered. No one listened. The next year, I. W. Gray, a Seattle businessman, took up land between Fish Trap and Anahim Lake. It isn't known who was expected to buy the land, nor why anyone would want to, but nevertheless PWD gave him a $5000 grant to build a wheel-capacity road from Tatla Lake west. During the road construction, Gray caught pneumonia and died in camp. His body was carried out on an improvised stretcher, then taken by wagon to Graham's where a temporary coffin was made for the car trip to Ashcroft. When the vehicle carrying the body broke down at 74 Mile House, a PWD truck took the coffin on to Ashcroft where it arrived four days after leaving Tatla Lake. Gray's son Paul finished the road, but the PWD inspector rejected everything except the McClinchy bridge, a sixty-foot span with "good crib abutments." He said the road work was done by "men unaccustomed to road building." Young Gray was paid after some argument, but PWD billed the estate $25 for trucking his father's body to Ashcroft. The estate protested that too, and the deputy minister finally agreed that as an employee of sorts, Gray Sr. was entitled to a free ride.

Domas Squinas didn't fare so well. Squinas, whose family for many years were the sole residents of what is now Anahim Lake, sent PWD a bill for $350 for bridges and trails he built in the same area at the same time. Officials said there was no record of his work—it didn't occur to them he might have built Gray's road. The McClinchy bridge was a blessing to residents; the river was difficult to ford because horses couldn't get their footing on the slimy rock bottom.

At Tatla Lake, Bob Graham reigned as the cattle baron of west Chilcotin. He acquired a post office in 1913, and in the late 1920s he started a store. The closest merchant at the time was Bob Pyper, an ex-policeman, at Chilanko Forks. Chilanko was much too far away for the West Chilcotin folk who considered it a chore to get to Tatla Lake. They kept getting supplies from Graham, and he decided to have the name as well as the game. Graham also had his own sawmill, powered by a steam engine taken into the country with great difficulty. He used his own lumber to build the "big" house, and then Mrs. Graham began taking paying guests, partly for store and post office customers, but also because Tatla was at the end of the navigable road and a natural stopping place for travellers.

The Grahams had three children and the Knolls at Chilanko had five, but families were in short supply past Redstone until the 1920s when a dozen new settlers arrived from the USA. Their arrival worried the old-timers in west Chilcotin, who feared the country was getting crowded. There was a clutch of bachelors among the new lot, but Andy and Hattie Holte had three children, Cyrus and Phyllis Bryant four. It was enough for a school at Tatla. The establishment of the school set the scene for another west Chilcotin milestone. Sam Colwell married one of the teachers, an event that led to the care and comfort of Chilcotin Road travellers for the next thirty years. A big woman with a big heart, Gwen Colwell was a magnificent cook. No one got away from her without being fed. The Colwells were such good company people would have stopped there to visit anyway, but it is safe to say Sam and Gwen gave away more food than they ever ate themselves. Gwen ran the telephone switchboard for years. Travelling dignitaries made a habit of stopping to see if there were any messages, knowing full well they would be invited in for a meal or a snack.

Among the single new arrivals to West Chilcotin were Tom Chignell, Lester Dorsey and Leonard Butler. Tall, slim, soft-spoken, Chignell was the son of an English doctor. His father knew R. C. Cotton, and after World War One Tom wrote to him asking about the Cariboo. Cotton wired "Come," so Tom came. He worked for Cotton, Becher and Bob Graham before starting his own place a few miles west of Graham's. Dorsey, Butler, Holte and Bryant came from the US looking for the good life with more optimism than was warranted. Butler settled south of Tatla, the other three ended up in the Anahim Lake area. These men walked well with the Drummer. They preferred freedom to progress, and Dorsey, Holte and Bryant knew the Chilcotin Two-Step before they ever came to the country.

The Bryants lived in the Soda Creek area when they first came to BC, so Cyrus certainly knew all about Chilcotin winters. Nevertheless, he didn't get around to moving his family west until mid-December. There was over two feet of snow on the ground when they left Riske Creek in two wagons and a buckboard. Their possessions included Phyllis's treasured piano, which had travelled with her from the US. The older children, Jane, Caroline and Alfred, aged eight, seven and six, rode saddle horse, cowboying a dozen head of cattle and six or seven extra horses. Bunch, the baby, rode in the wagon with Mrs. Bryant.

The first day out the temperature dropped to -30°F, or as Chilcotin people simply said, 30 below. The horses had a tough time pulling the heavy wagons through the snow. The family camped in tents. Phyllis had cooked a big mulligan stew, which froze, so they thawed chunks of it over the open campfire. The third day out Mrs. Norman Lee invited them to stay overnight. The thought of a bed was heaven to Phyllis. The animals probably

*Chilcotin Hotel, Alexis Creek. When Alex Marshall built this hotel in the late 1920s, it was a one-storey stopping place. Gus Jakel bought it in 1934 and added a second storey and a beer parlour, which for thirty years was the only liquor outlet between Williams Lake and the Pacific. Gus installed the first water system in Alexis Creek and the hotel boasted running water and inside plumbing, both firsts for Chilcotin. The original hotel burned in the 1970s. It was replaced by a modern building. (BCARS 84365)*

*Modern times. Hugh Bayliff at the wheel, and his wife Gertrude in the back along with neighbour Tommy Young, setting off for a tour in the 1920s. Bayliff drove the car until it dropped. It ended its life providing power for a small sawmill.*

*Becher's Prairie. Along with running a store, post office and stopping place, Tom Lee ran a taxi between Williams Lake and Tatla Lake in the 1930s. He also hauled freight for the store, which meant tackling the road in all conditions. Here he is on Becher's Prairie on a good day—the road hasn't drifted. When it did, motorists had to detour around the drift, which exposed them to another problem: the large rocks poking up all over the Prairie waiting unseen beneath the snow.*

*Pry poling. When a vehicle got hopelessly mired in the mud, there was always a tree handy to chop down, or a pole left by an earlier victim. The operation took at least two people: one to sit on the end of the pole to jack up the vehicle; the other to poke brush, sticks, rocks or whatever under the wheel to give traction. Hodgsons' drivers were masters of the pry pole; they had to be. In this photograph, Jack Hodgson and Fred Linder pry their 1935 Ford Two Ton out of a mud hole on the Chilanko Forks route below Bear Head Hill. Their passenger got the job of sitting on the pole, and he had to climb up on it to do so.*

appreciated the stopover too: Cyrus didn't have hay for them and the snow was too deep for browsing. Cyrus left a kerosene lamp burning in the wagon with the vegetables to keep them from freezing, and during the night it tipped over. Everything in the wagon was lost in the resulting fire—linen, clothing, the piano. The family spent five more long, wearying, bone-chilling days on the road. People they met along the way gave them what they could in the way of food, clothing and household goods. Phyllis Bryant never really got over the trip. Chilcotin frontiersmen bragged they were "men who were Men and their women were proud of them," but the Bryant trip was the kind of thing that gave west Chilcotin the reputation for being fine for men but hell on women and horses.

With the exception of Tom Chignell, who had an engineering background, the Anahim settlers were not into machines. They were horse people. They had no money for cars anyway. Few vehicles ventured beyond Tatla. No one agrees who drove the first car to Anahim. Bob Melville, the forest ranger stationed at Alexis Creek, drove a pickup truck to Anahim Lake in 1924 in the fall, on the frost. Tommy Hodgson made a trip that year, or the year before, in a car, in December. Domas Squinas's son Thomas and old Capoose had a 1924 Chev one-ton in 1928. Alex Marshall delivered it to them. They drove it all over the place. It ended its life as a horse-drawn Bennett buggy—probably the first one in Chilcotin—because they couldn't get parts for repairs.

Anahim people didn't fuss about the roads, but Kleena Kleene settlers did. Early in 1925 the ten landowners in the area signed a petition asking for an engineered road to bypass Bob Graham's. They said road work was always done to his benefit and they wanted "no more grafting by acquisitive ranchers." They also objected to the four gates Graham had across the road. Kleena Kleene was in the Prince Rupert District, and officials there asked Cariboo district engineer Evan Jones to investigate. He made a number of trips west that summer. He didn't think much of the road, but he suspected most of the fuss was caused by cabin fever, a mysterious malaise that strikes Chilcotin folk from time to time. It's an outbreak of ill humour caused by long periods of bad weather, isolation and frustration. Entire communities come down with it, and it always has a target—in this case, Bob Graham. Jones didn't reroute the road, but he asked Graham to remove two gates.

Over the years, the road between Chilanko Forks and Tatla Lake changed a dozen times because no one could agree how to get around a big hill which had Loon Lake on one side and a deep gulch on the other. There were great rows between Graham, who had the best political connections, Arthur Knoll, who was pig-headed, to put it mildly, and Bob Pyper, who could be a miserable old coot. Benny Franklin built the original road on the south, or Loon Lake, side of the hill. He got around the steep bluff by replacing the wheels on the high side of the wagon with sleigh runners. He'd leave the runners on the hill, and reverse the process when he returned. Not everybody cared to do this and the road moved back and forth as people tried different solutions. After one move bypassed Pyper, his store burned down. Dan Pocket, a teamster who hauled freight around the country, went to jail for the arson and died there. A few old-timers thought the fire was convenient for Pyper, who moved to the new road. The road eventually ended up on the north side, skirting the gulch.

James Mackill was forty-five, older than most, when he went to Chilcotin. His family were in their teens. Mackill was a wiry, five-foot-ten man with a droopy red moustache. Educated in Edinburgh, he spent eight years on the prairies with the Mounted Police. He then returned to Scotland, didn't like it and came back to work for Gang Ranch at Beaumont Meadows, at Riske Creek. He married Anne Swanson, the daughter of early settlers. Her mother's first husband was William Manning, the first Chilcotin settler, who was killed at Puntzi in 1864. The Mackills moved to Soda Creek to farm, and raised three children. James worked as road foreman for ten years but the work wasn't steady. The Bryants had been neighbours and when Cyrus wrote raving about Chilcotin, James went to look with the idea of starting a hunting lodge. When he first saw One Eye Lake, the afternoon sun was dancing on the water and the wind riffles sparkled like sequins. On the far shore the dark piney hills, dappled and shimmering with pale green aspen, humped their backs against the impeccably blue sky. "It's called One Eye after an Indian chief," Cyrus explained. "He died a dozen years or so ago. His canoe is still up on the bank up there." Cyrus gestured to the west. "His daughter married George Turner and they live down the Kleena Kleene." Mackill didn't say anything. "There's all kinds of game," Bryant continued. "It used to be caribou country, but they've gone to the hills and now moose are coming in. Good trapping here too. Everyone makes a few dollars on furs."

Mackill was hooked. He went home, sold his cattle, bought more horses and applied for the land. The family moved to One Eye Lake after breakup in 1926. They built a cabin with green logs and that winter, when the logs dried, the wind and frost scrabbled in through the cracks. Mrs. Mackill scrounged all the newspapers and magazines she could get her hands on to put on the walls for insulation. She used flour and water paste, and the mice liked it. James and the boys—Jack, age seventeen, and Clarence, age fifteen—wore out the handles on two crosscut saws trying to cut enough wood to keep the cabin warm. They never succeeded. The three men spent a lot of time trapping, but prices were low and it was a long, hard winter. The next summer the Mackills built a lodge, and by the fall of 1927 they were ready for customers. October was wet, followed by another long, hard winter, and before the Mackills or any of the new settlers had a chance to establish themselves, the Depression came along.

Kleena Kleene settlers got scrapping about the road again in 1929, and this time Mackill was in the fray. Whenever the river flooded, and it frequently did, the road by McClinchy's went under water, forcing everyone living on the west side to canoe down One Eye Lake to get to Graham's. Even though they seldom made the trip, the water detour annoyed the bachelors. They began agitating for the road to go north of the Lake, bypassing Mackill. Jones paid another visit. He agreed the existing road wasn't worth fixing, but he wouldn't see Mackill stranded either. He repeated the recommendation that PWD engineer Fal had made fifteen years earlier—divert the road away from the river before any more money was spent on it. Nothing happened about that, though Mackill wrote so many letters that Prince Rupert opened a special file on him, and finally hired him as permanent part-time road foreman for west Chilcotin. The road work, guiding and post office kept Mackill busy but he always had time for community affairs, and he was known as "the mayor" of Kleena Kleene.

While life went on more or less as usual in Chilcotin, the 1920s saw a major change in Bella Coola. In 1924, three floods hit one after another and tore the townsite to pieces. When Premier T. D. Pattullo inspected the destruction, he was so appalled he decreed the entire town should move to higher ground on the south side of the river. The government bought property from the widowed Mrs. John Clayton, surveyed the lots, and traded them for land on the north side. Residents grubbed out trees and underbrush on the new lots, dismantled their houses, moved them from the sunny side of the river, and rebuilt them on the gloomy side, in the mountain shadows. It took three years to complete the move. In spite of all this, people kept pushing for a road out. Perhaps to cheer them up, the government reacted, and in 1929 hired Colonel J. M. Rolston to survey a road to Chilcotin. Rolston knew the territory: he'd surveyed for the railway in 1912–13, and "inspected" for a road in 1922. Now he had a contract for $25,860 to try again. The road was supposed to be built in 1930, but the money disappeared when Wall Street crashed in 1929, and the road dreams crashed again.

# CHAPTER 10

# It's a Bugger, Mrs. Becher

*Is too bad this depression came along same time as these hard times.*
—George Meyers, Stone

The 1930s were dismal times in Chilcotin. The winds forgot to whisper, the pines forgot to sigh, the rain forgot to fall, the grass forgot to grow. There were two summers of drought, no rain at all. The countryside shrivelled. Day after day the sun blistered down, sucking every drop of water until the creeks were dry, the lakes were low, the land was dust. The swamp grass was so short mice couldn't hide in it. Lower Chilcotin was plagued with grasshoppers so thick they clogged cars' radiators. A forest fire in the Anahim Lake/Towdystan area burned from spring to freeze-up, blanketing the entire country in smoke. The winters were some of the coldest on record.

Cattle buyers roamed the country. They paid two cents a pound for large steers, then offered to loan the settlers money. With no hay to speak of, some ranchers drove their herds to the Williams Lake stockyards only to find prices were $21 a head lower than the lowest price in the 1920s. Everyone ended up in the hole. Things got worse faster than they got better. The winter of 1934 was one of the coldest ever recorded in Chilcotin. Temperatures dropped to -70°F and stayed there for weeks. Game animals died by the hundreds. Cattle died too. Ranch horses pulled the gunny sack chinking from between the cabin logs and ate it.

Hodgson's driver Fred Linder stopped at Becher House for a bite to eat one day during this worst of times, just between breakup and spring, when the road was awful. Linder was weary. "How is everything with you, Mr. Linder?" Mrs. Becher asked. "It's a bugger, Mrs. Becher," he replied. The old lady was startled, but "It's a bugger, Mrs. Becher" became a Chilcotin catchword.

Settlers did whatever they could do to make a dollar. Almost everyone did some trapping. The government paid a bounty because there were so many horses on the ranges, so ranchers went on wild horse roundups: in one year 6000 horses were caught. Some men went prospecting, and there was a flurry of mining activity in the Tatlayoko and Whitewater areas. Road work came into its own as settlers worked off their land taxes (the more affluent hired someone to do it) and made a few dollars besides. Everyone who had horses tried to get their teams hired on. The road jobs made the difference between being dirt poor and getting by; it paid more than a few ranchers' bank loans and kept settlers in flour and beans.

The early road crews were called the push and grunt crews because their main equipment was "armstrong"—all they had was their two arms and brute strength to wrestle rocks, shovel gravel, grub, slash brush and operate their two major pieces of machinery, the fresno and the pull grader. Road work was done between spring and freeze-up although a few settlers were hired to "turn water"—divert runoff—during breakup. After James Mackill became foreman at Kleena Kleene, Chilcotin was divvied into four parts with Mackill at the west end (Tatla Lake to Anahim, more or less), Dave Fraser (Redstone to Tatla), Bill Bliss (Redstone to the east end of Hance's Timber) and Dave Chesney (Riske Creek). Bliss and Chesney had the most work because they had the heaviest traffic. Fraser was a wee Ontario man who came to Chilcotin to look after race horses for one of the settlers who didn't hear the Drummer. Fraser did. He took up property west of Redstone and was a star boarder at Stuart's, where he helped around the place. He had stomach trouble and when it was bothering him he wouldn't talk. If anyone spoke to him, he'd whistle. Chesney, the Riske Creek foreman for fifteen years, was a fair but tough boss. He had a crew fencing in late fall once when the weather turned mean. It was a case of do or die because if the work wasn't done then, it never would be, the money would be spent somewhere else. The men had difficulty digging post holes in the frozen ground and they were frozen too, so they did a lot of bellyaching. "You hired on to be tough shits, don't complain now," Chesney told them. The remark became another Chilcotin catchphrase.

The crews didn't have much to work with. Mackill and Fraser shared a pull grader, and in the fall of 1935, Mackill's total inventory was one seven-year-old tent, two long-handled shovels, one sack of blacksmith coal, one gravel wagon with box, and two 6-inch clevises. Because road jobs were in high demand, the four Chilcotin Road foremen were under a lot of pressure from their relatives, friends and neighbours. The foremen did their best. "If you have any money available for this section near Towdystan, I would like to suggest to you it be spent in the fall gravelling the same as last fall," Mackill wrote the district engineer in 1934. "It does better work and it gives the settlers a chance to have a few dollars for supplies to go into winter."

Road work didn't save Fred Becher. But the Depression can't be blamed for all his problems: it was progress—the arrival of the PGE Railway in Williams Lake in 1919—that defeated him. Chilcotin residents could do all their business and get all their supplies in the new centre, which was comparatively handy compared to Ashcroft. They had no reason to stop at Riske Creek, they could stay overnight "in town" on trips out. When Prohibition closed the saloon a short time later, Becher lost his best moneymaker. White-haired, stooped, nearly blind, harried by his debts and plagued by poor health, he was gallant to the end. When he died in 1936, the shelves on his store were bare, but almost everyone in Chilcotin went to his funeral to pay their respects.

While the glory days at Becher's faded, downtown Alexis Creek was developing into Chilcotin's government centre. The hospital, post office, school, telephone office, police station and forestry station were located at "the Creek." The businesses, T. C. Lee's store and stopping place and Alex Marshall's garage and stopping place, were doing well too, when tragedy struck the Marshalls. One spring weekend in 1933, Marshall, his wife and a neighbour were sailing on Puntzi Lake when a sudden storm toppled them. Mrs. Marshall clung to the boat as it was buffetted down the lake, but the men were lost. Peter Hunlin, from Redstone Reserve, was camped at the end of the lake, and he rescued her. He wrapped her in blankets, gave her hot tea, then rode for help.

Ironically, Hunlin died on Puntzi's shores fifty-four years later.

He was fishing at his favourite spot when he fell into the chilly water. There was no one to save him from hypothermia.

Gus Jakel bought the Marshall place. A Russian-born German who came to Canada as a child, Jakel arrived in the Cariboo in the 1920s, worked here and there, and had a freight truck and garage at Norman Lee's before he moved to Alexis Creek. His oldest son Gordon, like most Chilcotin youth, got his first road job at an early age. He was fifteen when Bill Bliss hired him to load gravel wagons. Gravelling was an endless job: one summer's gravel disappeared in the next breakup's mud. The wagons had loose boards on the bottom that were pulled out to dump the gravel on the road where Bliss and Roy Haines spread it with the pull grader. There were five or six wagons working. Gordon was working with two old pros who knew how to conserve their strength. He nearly killed himself keeping up with them. He was paid $3.20 a day for an eight-hour day, six days a week. A man and team received $6.40, and everyone paid 50 cents a day for room and board. Gordon thought it was big money.

The road crew's few bits of machinery were not user-friendly. The scraper, or fresno, was an ill-natured piece of equipment. The big, heavy, metal scoop with two handles on the solid end was used to move dirt. Two men worked it, both on foot, the first a teamster with two horses to pull the contraption, the second man controlling the scoop. The scooper held the fresno's handles in position to scrape up dirt, tipped it back when it was full, held it while the horses pulled it to the side of the road, then tipped it forward to dump the dirt. The summer Fred Brink was fifteen, Mackill put him on a fresno with Frank Obermeir, a big man with a small sense of humour. The operation was a disaster. Even big, experienced men had trouble with fresnos and Fred was neither. The scoop kept digging itself into the ground and flipping Fred over top of it. Every time something went wrong the horses spooked and Obermeir lost his temper. Finally Mackill put Fred to work picking rocks. Obermeir didn't last long in Chilcotin.

A pull grader with an underbelly blade replaced Bliss's multi-blade drag in the late 1930s. Some time later he got a small Holt tractor to pull it. Bliss was a master graderman but the tractor nearly did him in. The graders, known as puddle swipers, were tricky. Operators had to be skilled, agile and strong. They had to work the two wheels controlling the blade while the grader bounced along the road trying to throw them off balance. They had no control over the forward motion of the machine. Whenever the blade hit a solid rock, which it frequently did, the back end of the grader reared up, and if the operator wasn't fast enough to jump off, the machine chucked him. Even if the teamster was daydreaming, horses would stop as soon as something went wrong, but the grader could upset and toss the graderman before the tractor driver knew it. Bliss had a tough time on some particularly rough ground in the mid-1940s when Roy Haines was driving tractor for him. Einar Nordberg was picking rocks for them, and he spent a lot of time picking Bliss off the road too. Then in his seventies, Bliss retired soon after. He'd spent twenty-five years on the road.

Charles (Bud) Barlowe, who was raised at Soda Creek, operated the first power grader in the Cariboo District. Sometimes a swamper rode with him but mostly he was on his own. A truck would leave fuel caches. If the grader broke down, Barlowe fixed it. If the grader broke through a wooden culvert, he fixed that too, once he got the machine out of the hole. Paul Niquidet of Williams Lake always remembered a Chilcotin snowploughing trip he made with Barlowe. The snow was deep on Becher's Prairie. "The only way to get the grader through the snowdrifts is to bunt and back up, bunt and back up," Barlowe warned Niquidet. "We'll spend hours going nowhere." But this time the snow crust was so hard the grader rode on top of it. When it broke through, Niquidet learned why he was there—to shovel. And shovel and shovel. It was -25°F when they stopped for the night at Becher House. The bedrooms were icy but Barlow was an old hand. "Pinch the comforters from the empty rooms," he told Niquidet.

They were expected to plough nonstop from Alexis Creek to Chilanko and back. They'd eaten their lunches before they reached Stuart's so they bought a couple of chocolate bars. Someone had ploughed the road just past Redstone, so they turned back. It kept getting colder and the grader was so stiff the wheels spun every time Barlowe put the blade down. When Niquidet saw the lights of Alexis Creek he thought they were home free, but they couldn't get around a tight corner because the stiff blade kept hooking the frozen snowbank and pulling the grader into it. They got to Jakel's at 5:00 a.m. after thirty-six hours without sleep. Their last meal had been the chocolate bar fourteen hours earlier. As he was parking the grader, Barlowe, who was coming down with the flu, accidentally shmucked Jakel's brand new hotel sign. The temperature sat at -50°F for three days. When Barlowe tried to get the grader going on the fourth day, the blade snapped. Williams Lake finally sent a truck out to rescue them. Barlowe was off work sick for six weeks. Niquidet didn't work for PWD again, and he has poor memories of Chilcotin in winter.

Although the road foremen were supposed to suggest what work needed to be done, PWD responded with dispatch whenever an influential landowner squawked. When Bob Graham of Tatla Lake wanted $150 to pull stumps out with the frost, he got it. When Norman Lee's son Dan wanted Lee's Hill (near his place) upgraded, Bliss was instructed to do it. Some settlers were very direct. Arthur Knoll bought a steam engine in 1929 but he couldn't get it home because the road around the bluff in Bull Canyon was too narrow. When Bliss explained he had no authority to widen the road, Knoll said he'd take the engine through anyway and if it got stuck and blocked the road that would be too damn bad. Bliss got the money. It took a month to "blow" the rock and clear the road.

Some requests from not-so-large landowners were heeded. In July 1937, Pete McCormick wanted to fix the road between his place and Mackill's so he could get his mail without scraping the bottom of his car. "Remember meeting me at a bluff?" he wrote R. W. Ramsay, a district engineer. "I was waiting with two horses to haul a party coming up the road through a mud hole. Well, that bluff caved in and on a wet day it's dangerous. Don't you think something ought to be did?" Ramsay gave McCormick four days' work and something was did.

In spite of, or perhaps because of the hard times, there was the odd conflict between ranchers and PWD. Mackill could never get a crew when Bob Graham was working his sawmill because Graham paid better wages than the government. Cozins Spencer,

*Thomas Squinas. Domas Squinas had two sons, Louis and Thomas, and a daughter, Balonic. Thomas, pictured here in the 1930s with young Darcy Christensen, was a west Chilcotin legend. As a trapper and guide he had no equal, and he knew the Anahim Lake area as no one else before or since. Like his father, he straddled the line between the Native and non-Native communities and was liked and respected by all. Darcy, shown here with his first coyote–snared with Thomas's help–ranched at Anahim, then took over the store when his parents, Andy and Dorothy, retired. Known as the Flying Fur Buyer, he had his own aircraft and flew directly to the remote traplines to service them.*

*A Chilcotin welcome. Eve and Joy Dickinson went to Chilcotin in 1935 to work for K. B. Moore at Tatlayoko. Moore ranched and provided lodging for miners who were busy in the area. The sisters took the PGE from Vancouver to Williams Lake, then hitched a ride with one of Hodgson's trucks to Tatla Lake. Along the way the truck foundered in a hole near what is now the Chilancoh Rosa Ranch past Redstone. Eve, who married Tom Chignell, is in the cab. The men (l. to r.) are Jack Hodgson, Fred Linder and Bert Lehman. Joy took the photograph.*

the millionaire owner of Chilco Ranch, couldn't get a hay crew because PWD paid more than he did. He was so mad when he thought Bliss was raiding his workers he wired the premier demanding Bliss stop doing road work until Chilco's hay was harvested. The premier refused. Spencer, who made his money in the movie industry, bought Chilco Ranch in 1923. He was responsible for Chilcotin's biggest scandal seven years later when he shot two of his ranch hands, killing one, and disappeared. His body was found weeks later in the river. His widow subsequently married the ranch carpenter and sold Chilco. These events rocked Chilcotin.

People in west Chilcotin had the toughest time during the Depression. Summers were too short to grow a garden, and the true frontiersmen weren't much for milk cows, pigs or chickens. They couldn't make money doing road work because there was no road.

People got along as best they could, trapping, living on moose meat and bannock. "You were all right as long as you had salt," according to Alfred Bryant. The Bryants moved from Tatla Lake to ranch at Anahim Lake but they had a tough time of it. One winter Alfred and his dad Cyrus had only one pair of boots between them. They took turns: one would wear the boots, the other would stuff his socks with hay. "People with money wore moccasins under rubbers," Alfred said.

West Chilcotin saw some changes during the Depression. Frank Render sold his Kleena Kleene property to Carl Brink of Bella Coola, and moved to Lily Lake, near Towdystan. The Brinks were back and forth until the boys, Victor and Fred, finished school and moved up to run the ranch. Lester Dorsey and a partner, Austin Hallowes, bought Schilling's Three Circle Ranch, but Hallowes later disappeared after a beef drive. That mystery is unsolved to this day. The Andy Holte family moved to Anahim too, and there was the usual assortment of bachelors living in the bush here and there. The biggest change came about when Andy Christensen, a colonist's son, and his wife Dorothy, John Clayton's daughter, moved up from Bella Coola in the early 1930s to ranch at Clesspocket. Andy looked and acted more like a big city banker than a Chilcotin rancher, but he knew the Drummer

well. He ran the ranch and the store, traded furs, was justice of
the peace for thirty years, and in general was the "mayor" of
Anahim. The Christensens were a godsend to Anahim.
Clesspocket became the main source of jobs as Andy usually
needed cowboys or hay crews. He believed in progress, up to a
point. Clesspocket had a telephone, a generating plant and a
bulldozer, but Andy never did put running water or plumbing in
his home.

Like west Chilcotin settlers, the Bella Coola colonists could rely
on road work for a few extra dollars now and then, but unlike
Chilcotin, politics played a major role in the process.

Foremen's jobs were political plums in the valley, and whenever the government changed, the foremen changed. Tom
Draney and Ivor Nygaard took turns at one foreman's job. They
each had big families and it was no joke to either of them to be
out of work, not because they weren't doing a good job, but
because they voted the wrong way. Like the Chilcotin Road, the
quality of the valley road diminished as the population thinned.
It ended abruptly at a little cabin at Stuie, where the Atnarko
and Talchako Rivers meet to form the Bella Coola. The valley
was divvied into three "foreman" sections. The crews shared
one set of equipment, a tractor, a pull grader, and two trucks.
Whoever got them first kept them; there were great fights
among the foremen, and among the residents of different parts
of the valley.

The Depression and hard times went together in Bella Coola,
too. In 1934, a huge flood put Hagensborg under water. Two
years later another one destroyed the swinging bridge between
the Nuxalk village and the townsite. It also took the water line,
so Indian Affairs officials moved the entire village across the river,
next to the townsite. Just as the colonists had done a dozen years
before, the Nuxalk people dismantled their houses on the sunny
side of the river and moved them over to the shady side, under
the mountains.

The colonists never stopped harping about a road "out," and
every few years someone in government would react. In 1935, the
government hired a surveyor, Ernest LaMarque, for 8 dollars a
day and expenses to relocate the Alexander Mackenzie Trail west

of Atnarko. He found the Palmer route to be unsuitable, but he thought he could find a good location. Nothing came of it.

The lodge, trapping, post office and road work kept Mackill busy, but he always had time for community affairs. He was the mayor of Kleena Kleene, and he chaired all the important meetings in west Chilcotin for years. In 1935 he bought a 1927 Dodge with a high clearance, and when road conditions were right he took guests to Nimpo and Anahim Lake. In 1942, when he was sixty-two, he went to Vancouver for an operation and didn't make it. His youngest son Clarence and wife Rocky came back to run the Lodge with Mrs. Mackill, and Clarence worked on the road.

CHAPTER 11

# The Road Followed His Wheeltracks

*. . . to Redstone OK except for clay section. From Redstone to Mile 100 (Tatla Lake) fair but impassable in some places in spring breakup. From Mile 152 (McClinchy) to Mile 182 (Mackills) spent $1000 but still several bad stretches. From McClinchy to Mile 198 (Towdystan) the road degenerates. It follows the line of least resistance through swamps, climbing to bench land. It should be relocated but circumstances don't warrant. Could be graveled. Towdystan to Mile 223 (Anahim Lake) same problem.*
—Public Works Annual Report 1940

The road west followed settlers, who followed the line of least resistance. There was one notable exception to this, a forty-mile chunk of road that followed a trucker.

Anahim Lake is on the western edge of Chilcotin. In the early 1930s, according to residents, Anahim might not have been at the edge of the world, "but you could almost see it from there." There wasn't much beyond it except the sunset and there wasn't much there except crows, moose and jackpine trees. It wasn't even Anahim Lake, it was Anaham Lake. The dozen or so settlers lived on swamp grass meadows far away from each other. There was no post office, school, store or central gathering place except Christensen's Clesspocket Ranch. The usable road from Williams Lake ended at Kleena Kleene post office. Nobody within fifty miles had a licensed motor vehicle. Settlers rode saddle horse to Kleena Kleene three or four times a year to get their mail. They went on to Tatla Lake if they absolutely had to. Bella Coola was their trading centre, and they brought their supplies over the mountains by packtrain a few times a year.

Stanley Dowling changed all that. An enterprising young man

in his early twenties, Stan was middling size, strong as a couple of bears, and tough as rawhide. He wasn't so much a dreamer of dreams as he was a doer of deeds. He wanted to own more than the sunsets, but he didn't mind adjusting his step to the Drummer if that's what he had to do.

Stan arrived in Chilcotin from Vancouver in 1932. His brother, who was working in Bella Coola, had invited him to visit. Things weren't so wonderful in the Depression-ridden city, so Stan quit his job and hopped a freight train to 100 Mile House. He hitched a ride to Lac La Hache with Julian Fry, a prominent Cariboo rancher. The next day he caught a ride to Williams Lake. Roy Haines took him to Lee's Corner the third day, and someone else took him to Alexis Creek. He walked twenty miles to Redstone. Andy and Hettie Stuart locked the store and drove him to Redstone rancheree, then he walked to Chilanko Forks. He spent two weeks at Bob Pyper's trading post sawing firewood, then caught a ride to Tatla Lake. Sam Colwell was there, and he took Stan to his place at Kleena Kleene.

Rides were nonexistent from there. Stan walked to Towdystan, following the telegraph trail on the north side of the McClinchy River. Andy Holte was renting Towdystan, Engebretsons were in Bella Coola, but when Stan arrived, Shorty King was minding the place while the Holtes were at a stampede at Stuie. When they came home, Stan and Shorty rode to Bella Coola where Stan wintered with his brother. In the spring he rented a quarter section at the foot of the Mackenzie Trail, near Burnt Bridge, for $25 a year. He had twenty-five head of cattle and six horses. He rented pasture and sold vegetables, but there wasn't much traffic and the place was no money maker. The next year he and Shorty packed all the machinery and drove the cattle to Anaham Lake. Shorty fed the animals on shares while Stan went to Vancouver to make a few dollars.

He had a plan. Everything people in Anaham needed—food, clothes, furniture, machinery—came to Bella Coola by boat, then over the mountains by packtrain. The process was time-consuming and costly, and packtrains couldn't climb the mountain in winter. Stan thought he'd truck supplies to Anaham directly from Vancouver, hauling furs, cattle, horses or whatever else there might be out of Anaham on the empty truck. When Anaham people heard his plan, they thought he was nuts. Packtrains were good enough, they said. Stan was getting too high-toned. Besides, how could a truck get in and out of Anaham when a wagon couldn't make it over the trail? They laughed all winter.

Stan thought he could do it. He made a deal to have his freight kept at Bert Lehman's place, because the six miles from there to Anaham really was impassable. He worked at Clesspocket that winter. In the spring he traded four two-year-old steers for a 1926 Moon car with a padded leather body and hydraulic brakes—a first. He walked to Redstone (120 miles), bought a battery from Stuart's, and carried it six miles to Ross's ranch to get the car. It needed brake fluid but the Alexis Creek Garage didn't have any because other vehicles had mechanical brakes. They tried a mixture of oil and gas which dissolved all the rubber in the braking system, so Stan drove to Vancouver with no brakes.

He traded the Moon for a 1933 Ford with a four-cylinder motor which was interchangeable with a V8 engine. They called it a BB model because it had two springs on the front, a first for Ford. He loaded it with three tons of freight. Thirty pack horses couldn't have carried it all. He had tools, machinery, some secondhand furniture, dry goods, all kinds of groceries, everything he could think of that somebody might buy. He didn't haul any grain because no one could afford to buy it—chickens ate moose meat and table scraps. He also carried a winch, chains, ropes, a cable and several jacks in case he had trouble on the home stretch.

Mother Nature had been unusually generous with rain that spring and she let loose a special deluge to greet Stan. Mackills, who recorded the weather for the government, had measured six and a half inches of rain in two days. It brought the snow down from the mountains and the whole country was awash. Stan sloshed by McClinchy Ranch, which was under water, and made it over the bridge at Brink's, which was floating, but he was stopped cold at the McClinchy River. The bridge had just been re-decked but the rushing river swept the planks away. Alerted by moccasin telegraph that Stan was on his way, Bert walked down to meet him. The old decking was piled by the road, so they just lay the old planks on the stringers. Stan drove over the roaring river with the decking kicking out behind him.

Lehman's place was two miles off the main road and there was no road in to it, so Bert guided him through the bush. Everyone in the country was there waiting to see what was on the truck, even the sceptics showed up. No one had any cash, so Stan's business was trade. Everybody had furs in winter, and the rest of the year they traded whatever they had. Stan piled up a lot of deals but one truckload lasted a long time.

He made his second trip in late fall. The Ford's transmission cracked on the 100 Mile hill coming home and it took ten days to get a new one shipped from Vancouver. (It came by train to Ashcroft, then in a seven-passenger car to 100 Mile House. The car ran three days a week to Williams Lake.) All was well on the Chilcotin Road until he hit fresh snow at Towdystan. The new snow covered the old track, and in the grey twilight of late afternoon, he couldn't see where the road was supposed to be. There were sleigh tracks, but they took shortcuts. When he followed them he got into some predicaments and had to back out. The sleigh turned in at Lehman's so he followed, got good and stuck, and had to walk in. Bert helped him rescue the truck in the morning. The people waiting stayed overnight so they could get their supplies.

Vehicles took a terrible beating in Chilcotin. Since Stan had to have something reliable, he went to Vancouver at Christmas each year to trade for a new truck. A 1935 six-cylinder Chevrolet Maple Leaf replaced the Ford. This was a special Canadian model, there weren't many of them. Stan had it equipped with mud and snow tires which were new on the market and were supposed to take the place of chains. They didn't. It was quite a letdown.

The stampedes at Stuie were popular, partly because Tommy Walker at Stuie Lodge had been a brewmaster in England and kept his hand in. He and his partner A. J. Arnold had a little cabin they called the church where the guys went to drink. The story was you couldn't drink Stuie dry, but every year people tried. There wasn't much social life in west Chilcotin. Few folks went to the Williams Lake Stampede because it was too hard to get there. In 1937 Stan decided to put on a stampede at Lehman's around the end of June, after the Williams Lake event. Of course it rained. Stan had a few passengers with him when he was bringing in supplies for the stampede and the road was so bad they stayed overnight at Kleena Kleene Ranch. After Pat McClinchy died, a Russian who went by the name of Nick Cassidy had the place. In the morning Cassidy followed Stan with his team. It was a good thing he did because the creek not far from Lehman's was high. When Stan went through it his muffler submerged and he went short of power. Cassidy's team helped him out.

Provincial policeman Bill Broughton from Alexis Creek and two RCMP officers arrived to make sure the stampede didn't get out of hand. Broughton's van couldn't get over the creek, so they left it and walked. The water was so deep it soaked the tobacco in Broughton's shirt pocket. The RCMP took their uniforms off before they took the plunge. The bucking chutes weren't finished in time so the chutemen blindfolded the saddle broncs and snubbed them up to the saddle horn. The would-be rider got in the saddle and the chutemen pulled off the blindfold and let the horse loose. The bareback horses had to be thrown. The riders got on as the horses got up. Nobody was seriously hurt.

The dance hall wasn't finished either, only four rounds of logs were up for the walls, but Billy Woods was there with his violin. Cassidy had one too, and they played for the dance. Woods was a lanky, dark-haired cowboy/prospector who wandered around the Chilcotin, sometimes working on ranches, always looking for the big strike. His right arm was crippled when he fell off a bicycle as a child, and he played the violin left-handed. It was always a mystery how someone so ornery could make such toe-tapping music. At least two generations of Chilcotin people learned to dance to his fiddling. There was no charge for the stampede or the dance, and everybody enjoyed themselves. The police said they enjoyed it too.

At Christmas Stan traded the Maple Leaf for a 1938 two-ton Ford. George Turner and his wife went to Vancouver with him. They started home New Year's Day but they only got as far as Choate (between Hope and Yale) because the Fraser Canyon was snowed in. It wasn't going to be opened until spring, so Stan loaded the truck on a flatcar, roped it down and blocked the wheels, and he and the Turners got in the cab. The train hooked on to the flatcar and away they went. They were warm in the cab, and they had food to munch on. When the train stopped at North Bend, they went to have coffee at the cafe, and the conductor caught them getting out of the truck. He wouldn't let them back in, he said it wasn't safe, so when the train pulled out they jumped into an empty boxcar. It was wickedly cold and uncomfortable. At Lytton they unloaded the truck and Stan paid the station master for hauling it. "I should charge the three of you for riding in the boxcar," he said. Stan didn't tell him the truck was loaded with freight.

One June in Williams Lake, just as the stampede there was ending, Stan met a little Austrian named Alfred Lagler who was running a few games of chance. His buddies had a hot dog stand, and they all wanted to go to the Anaham Stampede. Stan advised against it, but on his way home he came across Lagler a few miles east of Tatla Lake. Lagler was in a sorry state. He was driving a Chandler car with his tent and gear on a trunk rack. His exhaust pipe had caught the lot on fire. He had two fellows with him, so Stan loaded them and their belongings on his truck.

The McClinchy River, which was spilling over the road, trapped the hot dog outfit's Model A. Stan loaded them and their belongings on his truck too. The stampede was another big success.

Stan wanted to move his store to Anaham Lake proper, but there was the problem of the last five miles of road. No one else wanted it fixed, and Stan wanted to change the route anyway. His idea was to follow the telephone line which went straight from Nimpo Lake to Anaham. Most of the line was strung on live trees. The right-of-way was brushed out and everyone went that way saddle horse, but local wisdom said a truck wouldn't get through the mud holes. Public Works wasn't interested in doing anything about either road. Jim Mackill wasn't pushing it. Hunters and fishermen didn't dare take their vehicles beyond his place so he made a few dollars taking them to Nimpo Lake in his car, which had very high clearance.

Stan walked along the line and found a way around the mud holes. On his next trip out he picked up a husky young chap in Williams Lake and armed with an axe each (no chain saws in

Road grading, 1930s. In the early days horses pulled the graders, which spread gravel and evened out ruts and potholes. During the 1930s Public Works mechanized, introducing bulldozers and tractors to pull the graders. Bud Barlowe is doing the honours here on the Cat.

Hance's Timber. During the winter of 1944–45, the Canadian Armed Forces staged an elaborate manoeuvre called Operation Polar Bear at Anahim Lake and Bella Coola, to test various kinds of equipment, and men, under winter conditions. When district engineer O. R. Roberts learned that the Department of National Defence intended to take all that heavy machinery onto the Chilcotin Road, he feared disaster. He persuaded Public Works to negotiate an agreement that left DND responsible for any damage the Polar Bear exercise did to the road. Roberts took Before and After photographs. This one is an After, of Hance's Timber. DND coughed up the money for repairs, but by the time Roberts got it the season was too far gone to do the work. If Public Works money wasn't spent by the end of the year it was lost, so Hance's Timber and other stretches of the road stayed in disrepair into the 1950s.

those days) they built six miles of road in four days. "If you chop the trees close enough to the ground I can drive over the stumps," Stan explained to his helper. They had to pull a few big stumps out with the truck. The road crossed the Dean River at Fish Trap, but there was no need for a bridge because the river bottom was solid gravel and the river froze in the winter.

The Bowser family and widowed Mrs. Domas Squinas were the only residents at what is now the village of Anahim Lake. Bowsers lived at the Hudson's Bay post. He worked on the telephone line and they put up any travellers that happened along. A few years before some people named MacKenzie had had a trading post just east of Bowser's, but it burned down and they left. The basement hole was still there. Stan built his store on top of it to save the bother of digging another hole. He traded goods for logs and labour, and soon had a 20 x 20 foot log building housing the store and living quarters. That fall he had a contract to haul Christmas trees from Lee's Corner to Williams Lake and he ran into Lagler again. He hired him to run the store at Anaham. Alf was a sliver of a man, but he was wiry and tough. He had black hair and dark burning eyes. He tended to be accident prone and excitable (the one often led to the other), but he was a great storekeeper because he had an eagle eye for a good fur, and people liked and trusted him.

Nobody in Chilcotin worried much about land title until after they were settled. When the store was finished Stan built a house and then applied for the land. That's when he discovered he'd built everything on Hudson's Bay property. Expecting the worst, he wrote to the company asking what they wanted for the land. He was pleasantly surprised. "I can buy the whole damn town for $750," he told Alf.

Stan's new road served its purpose until winter. When the snow got too deep he left the truck at McCormick's place at Clearwater and came the rest of the way by sleigh. Baptiste Dester was with him. (Baptiste, the son of early Chilcoten trader George Dester, settled in west Chilcotin.) The Dean River wasn't frozen and they couldn't get across the gravel bottom. "What if we cut some trees and lay the poles across the gravel so the sleigh can slide across?" Stan suggested. It took some time to get this organized and it should have worked, but the poles bunched up under the sleigh's back bob. Stan had to get a pole and pry up the back of the sleigh while Baptiste drove the horses. On the next trip Stan crossed a bit farther down river and the ice broke. He fell in, got soaked, and had to walk the ten miles home in freezing weather.

"To hell with this, I need a bridge," he decided.

Public Works would not part with any money for a bridge because they didn't recognize the road, so Stan gave Sam Sulin a team of horses worth $75 to build a pole, or punchin bridge. Years later when PWD accepted Stan's changed road they gave him $70 for the bridge—20 days at $3.50 a day, the going wages when the bridge was built. The bridge was fine for vehicles but horses didn't like the punchin.

With the store open most of the time and the freight truck making regular trips, the only thing missing was a post office. People still rode to Kleena Kleene to get their mail. Whoever went picked up everybody's mail, bringing what they could tie behind the saddle. As often as not Stan picked up the mail, so he thought he might as well have a post office. The Anaham Stockman's Association backed his application. Whoever got the first post office in a community got to name the post office, and the community usually took the name as well.

Anaham Lake was named for Chief Anaham, one of the better known Tsilhqot'in chiefs. Historians believe Anaham (with many different spellings) was the name given traditional chiefs. According to Ulkatcho elder Benny Jack, Anaham was from Stuie. He used to go over the mountains to hunt muskrats (Anaham Lake rats are heavy, seal-like animals, the best in the Cariboo). Tsilhqot'ins went to Bella Coola to trade, and somewhere along the line Anaham fell in love with and married a woman from Chezacut. She liked to visit Bella Coola, but she didn't like living in the rain forest, so she persuaded Anaham to move up the hill. Members of his Band followed him and he became chief of the village of Nankutlun on what became Anaham's Lake. The smallpox epidemic in 1862 devastated Nankutlun. When the Indian department moved the Tsilhqot'ins east following the 1864 "war," Anaham and his people moved to what is now Anaham Reserve near Alexis Creek.

Stan could have called the place whatever he pleased. Bowser's telephone was listed at Dean River, Stan's phone was listed at Dowling. He decided to stick with Anaham but he changed the spelling to Anahim so it wouldn't be confused with Anaham Reserve.

Since the post office didn't pay much, Stan put in a bid of $500 to haul the mail from Kleena Kleene every other week. He thought he'd haul the mail along with the freight, but he didn't get the contract. Sam Gladhill, a farmer at Canoe Crossing, got it. Gladhill had the mail contract to Atnarko, about forty miles downhill from Anahim, and he must have thought he'd just keep going. No one packed over the mountains in winter, the snow was too deep. Gladhill intended to snowshoe, but after he got the contract he backed out. The post office offered it to Stan for the same price. He'd been hauling the mail for nothing anyway so rather than delay any longer he took it. As soon as he got the contract the mail began to grow. Eaton's and Simpson's and other mail-order houses sent sacks full of catalogues. Parcel post was cheap—65 cents for twenty-five pounds anywhere in BC—and it didn't take long for Anahim residents to get on to a good thing. Andy Christensen started getting dry goods mailed in for his store at Clesspocket. George Cahoose, a Southern Carrier who had a store up the Dean River, began getting his groceries mailed in too. Johnny Weldon, who lived on the Precipice Trail, once got two 24-pound sacks of flour in the mail from Woodwards. Each sack had a piece of paper wrapped around it and came, first class mail, in a bag addressed to Anahim Lake. Stan loaded all this stuff on the truck at Kleena Kleene and off again at Anahim, and he suspected some of the parcels that came from the stores' mailing rooms were more than 25 pounds. He weighed one from Eaton's and it was 33 pounds. "I'd like to have seen Gladhill pack these loads over the mountains on snowshoes," Stan said.

Stan stocked all kinds of things in the store. Big sellers were items with premiums inside—boxes of soap powder with dish towels, three pounds of coffee with a cup, rolled oats with a bowl. He bought flour in calico sacks, and the women made underwear and dresses out of them. Stan also had the Massey-Harris (later Massey-Ferguson) dealership for thirty years. In the early days the company carried steel and rubber tired wagons, three different sizes of horse sleighs, kitchen stoves and heaters, dump and side delivery horse rakes, and all kinds of farm machinery.

When Stan moved to Anahim, the stampede moved too. He had bucking corrals built near the store. The first year they ran the horse races down the main street of the village. Later they built a proper track. The dances were held in his truck garage where the walls would shudder and bulge as the guests hopped up and down to the music.

Stan liked to go into winter with a new vehicle on warranty, and he usually had to use it. City mechanics couldn't imagine what he'd been doing to the vehicles. In 1940 he bought a three-ton with a two-speed axle. It had eight gears instead of four. That winter he wore out the case hardening on low gear, so he went to Vancouver to get the transmission overhauled on the warranty.

"There isn't any way it could have worn out so soon," the mechanic insisted when Stan arrived with his problem. "How much do you drive in low gear anyway?"

"One trip I was in low gear for six days," Stan replied. The mechanic honoured the warranty.

Another truck that did yeoman service was a 1939 two-ton 4-wheel-drive Ford. A standard Ford with a Mormon Harrington drive, it had been ordered by the army but never used. It was unique because it had no lever for the 4-wheel drive. When a rear wheel started to spin, it would kick into 4-wheel drive. Step on the gas, it would go back into 2-wheel. Stan paid $1000 for it. With four sets of chains on and all the cross links each chain would hold, it would go anywhere—very slowly. In deep snow its top speed was about three miles an hour. Riders would pass Stan on his way home and they'd go ahead to Anahim to tell everyone he was coming.

Stan's "six wheeler" was the truck everyone talked about. He bought it just after the war. It was an army truck made by Reo, a tandem with front-wheel drive. Stan put in a lower-geared transmission, added a freight box and painted the whole thing red. It was a classy looking outfit, but in winter he had to take all four outside wheels off or it would sit and spin in the snow.

# CHAPTER 12

# Winter Wonderland

*SNOW REMOVAL CARIBOO AND LILLOOET DISTRICTS–
the Public will please take notice no snow ploughing will be done
on any of the following roads: Likely Road, Horsefly Road, Canim
Lake road, 93 Mile Road, Lone Butte Road, Springhouse Road;
Chilcotin Road. All persons who are compelled to use these roads
should make arrangements now to have teams and sleighs available
for winter travel.*
–H. L. Hayne, Public Works Engineer, Williams Lake (1932
newspaper advertisement)

Settlers generally stayed home in winter because they had stock to feed, but truckers, mail contractors, police and longtime coroner Rene Hance tackled the road in all seasons. There were countless near misses but remarkably few disasters. Though the Drummer looked after his own, he put them to the test often enough, and everybody got caught out, even the most wary and experienced.

Hance and policeman Dan Weir had a close call when Riske Creek settler Jim Slee died in mid-December 1933. Gus Jakel was living at Norman Lee's at the time. He was coming home in a storm when he broke an axle at Riske Creek. He carried his freezables to Slee's, breaking trail in the deep snow, and got there about midnight. The storm had tired itself out by morning and the air was crackling clear when Gus and Slee went with a four-horse team to rescue the truck. Slee was building a new house near the spot where Gus was stuck, and when they stopped to get a chain, Gus suddenly realized Slee wasn't with him. He was slumped in the snow, dead. Gus couldn't lift him into the sleigh so he went to get Mrs. Slee. She was calm. She said her husband had had a ticky heart and she'd been afraid he'd die out in the bush where no one would ever find him. They phoned Weir and went back to get the body. While they were struggling to get it in the sleigh, a neighbour rode by. He'd fought with Slee some years before and didn't stop to help. Gus was still mad about that fifty years later.

Gus and Mrs. Slee got the body and themselves back to the house, but Weir didn't show up. Late in the afternoon Willard Bob from Meldrum Creek rode by and found Weir's vehicle stuck, with the coroner and policeman inside, barely conscious. The car's universal joint had broken bucking the drifts. The men knew someone would be looking for them sooner or later, and there was nothing they could do anyway. There was no way to find wood or make a fire in the deep snow, so although they knew the risk, they ran the engine every little while to warm up. Bob managed to revive them.

In 1934, the record cold winter, Tommy Hodgson carried "the mail must get through" motto to the limit on his Christmas run. He left on Friday, December 21 with the biggest mail he'd ever carried. The weather was fine, the road was good, and he reached Kleena Kleene on schedule Sunday night. The temperature took a header overnight, dropping to -38°F. It was snowing belligerently when he left early Monday morning, blasts of seething snow that blotted out the landscape and buried the road. He got to Bob Pyper's at Chilanko late that night, Christmas Eve. It wasn't a very silent night with the wind yapping and yowling around the cabin. Christmas morning the mercury was crouching at -50°F so he didn't try to leave. It was still bitterly cold Wednesday. Tommy spent all morning warming the truck and getting it started, and he managed to get to Alexis Creek late that night. The snow drilled down all the next day as he bulled his way to Hanceville. He left the truck there and went on with a team and sleigh. He met his driver coming with that week's mail near Slee's. They traded loads, and Tommy mushed back to Redstone. He got back to Hanceville on Saturday, December 29, and started home in the truck, which conked out on Borland Hill. He walked the last ten miles home.

Stan Dowling had his share of hellish trips. In January 1938, he spent eight days, going steady, getting home from Chilanko Forks, a distance of seventy-four miles. A blizzard had hammered the road into one solid drift. Stan's load included a new freight sleigh. He'd phoned ahead to have Andy Holte meet him at Chilanko with a four-horse team. Bill Graham was at Chilanko too. Rancher Ed Collett had broken his leg and Bill had taken him down to catch Hodgsons, who were hauling the mail in their cutter. Andy didn't think the team could haul the loaded sleigh so he and Bill broke trail while Stan bunted and shovelled his way to Tatla Lake in the truck. It took two days to get there, Stan catnapping in the cab. Anahim residents Alfred Bryant and Pan Phillips were at Graham's picking up some horses. They took some of the freight in Collett's sleigh and went ahead with the loose horses to break trail. Stan left the truck to winter at Graham's, and set out with Andy, the new sleigh, and the four-horse team.

The sky was grey and it was hard to tell if the snow was coming down or just blowing around. The drifts were like cement. The first day they made it to Cassidy's, thirteen miles, but one horse played out. Andy traded it to Cassidy for an unbroken mare which turned out to be too light to break through the drifts. Andy, who was a top teamster, sat on it to give it more weight. "She'll be broke by the time we get home," he promised.

It took all day to travel the ten miles to McCormick's. The horses were exhausted from lunging through the drifts. Alfred and Pan went ahead the next morning. Stan and Andy spent all the next day getting up McClinchy Hill. They camped at a spring on top, seven miles from McCormick's. Alfred had abandoned Collett's sleigh there. The campers had some protection from the wind but no feed for the horses.

The wind had sculpted dunes all over Caribou Flats and there was no track. Alfred and Pan had gone through the timber but the sleigh had to stay in the open. Stan and Andy camped out again, nine miles from Towdystan. The wind died during the night, and morning dawned breathless and glittering. A few miles up the road they ran out of snowdrifts and they staggered into Towdystan that night. Fed and rested, with fresh horses, Stan got home the next day.

Chilcotin weather can jump or fall 60 or 70 degrees Fahrenheit without warning. That's what it did when Frank Render died.

*Bill Graham. Bob and Margaret Graham's eldest son Bill spent all his life at Tatla Lake. Although his formal schooling ended at Grade Eight, Bill taught himself to be one of Chilcotin's most highly regarded mechanics. Bill also built a lot of roads. None is named for him but there is one in particular that should be. He not only built much of the Pyper Lake Hill road near Chilanko, he did the survey for it, riding horseback through the bush to pick the route. He must have had a good eye because when the highways ministry rebuilt the road, they didn't change it.*

Render was, by all accounts, a fine old man who never caused anyone any trouble while he was alive. Dead was a different matter. John Blatchford, a Provincial Policeman who was planning to move to Alexis Creek, was spending a few days there in March 1939, when the news of Render's death reached the incumbent officer, Bill Broughton. The two started out in Broughton's 1939 Ford panel truck, but the snow at Kleena Kleene was too deep for it. They borrowed a team and cutter from Jim Mackill, and got fresh horses at McCormick's and again at Towdystan. When they reached Render's log cabin at Lily Lake, they found him laid out in a homemade coffin. They fed their horses some of his hay and stayed overnight. They couldn't figure out how they were going to bury him in all that snow, but in the morning somebody rode in to say Render's sisters had wired money to Williams Lake; they wanted the body sent to Ontario. There was no room in the cutter for the coffin, so Broughton phoned Dowling to do the honours. When he and Blatchford got back to Alexis Creek the temperature was huddling around -40°F. They were greeted with the news that a man had frozen to death at Stone Reserve. It was too late to do anything that night. Instead they thawed out and caught some sleep.

Dowling was just in the door from his March trip when he got the message about Render. His truck was wintering at McCormick's. Alf Lagler said he'd fetch the body, and Stan could ride down and meet him at Towdystan the next day. It was a twenty-mile round trip for Lagler, who stayed overnight with the Bert Smiths at Morrison Meadow. Mrs. Smith decided to go to town too. She had to sit on the coffin because it took all the room in the sleigh.

A Chinook wind came along early that morning and when Mrs. Smith, the coffin and Stan left Towdystan, it was making quick work of the snow. Water was running everywhere. When they dropped into the McClinchy valley the road was bare and the horses dragged the sleigh on the ground five miles to McCormick's. They left for town the next day in the truck. The road was a sea of mud, it took them two days to slip and slop to town. The police gave Stan $50 for the trip. Render's sisters had no idea of the distance and hadn't sent enough money. "We figured you were coming in anyway," the police said. Stan didn't say anything, but he swore he'd never leave home again in a Chinook.

Blatchford and Broughton borrowed a team and sleigh for the trip to Stone. When they left Alexis Creek at 5:00 a.m. a warm wind was blowing. By the time they got to Stone, water was running all over the place. They put the body in the sleigh and headed home. Since the middle of the road was bare, they went from one side of it to the other to keep the sleigh on snow. Blatchford took the reins while Broughton caught a few winks. Blatchford was dozing along, swinging from one snowbank to the other, when he heard whooping and hollering behind him. It was a fellow from Stone on a saddle horse. He said the body had fallen off of the sleigh about a mile back. Blatchford turned around, retrieved the body from a large mud hole, and tied it in. When they got to the police station, near Lee's Corner, they had to wait for Rene Hance and the Alexis Creek doctor, Dr. Hallowes. "Hallowes was crude," Blatchford said. "He treated the body like a chunk of meat. It was my first experience with a body that wasn't in a coffin. It was quite a welcome to Chilcotin." The experience didn't scare him away, though. Blatchford left the police force after a few years and became a Chilcotin rancher and guide himself.

When Dowling, his wife Edna, and their tiny daughter left Anahim for a trip to town in the winter of 1942, the road was good but the weather was so cold the fuel line kept freezing. Stan had to keep stopping to unplug it, a cold and tiresome business, so they decided to stay over at Alexis Creek. The temperature plummeted to 50 below overnight, but Stan was determined to get to town. He unhooked the gas tank from under the seat, dumped the gas into a bucket, cleared the line, then poured the gas back in the tank, leaving the ice lumps in the bucket. This was all done outside: the truck rack was too high to get into Jakel's garage. With twenty gallons of fresh gas, plenty to get to Williams Lake, he set off that afternoon, leaving Edna and the baby at Alexis Creek.

There was a heavy snowfall just before the temperature dropped. The snow got deeper and deeper through Hance's

timber and it was hard going. Stan was happy to find the road
ploughed at Riske Creek. He was clipping right along when he
ran out of gas. The truck had gobbled more than its share getting
through the deep snow. He knew the road crew was at Becher's,
six miles back, so he starting hiking.

He covered his head with a blanket so he wouldn't frost his
lungs. At 50 below the cold slithers under collars and cuffs, trying
to get in and freeze your bones. It ices nostrils shut and burns
eyes. As he crunched along, Stan thanked his stars Edna and the
baby were safe at Alexis Creek. As he neared Becher's, he could
hear the throaty hum of the Public Works vehicles keeping themselves warm for morning.

It was nearing midnight when he stumbled into Becher's. He
was frozen stiff. The road crew knew better than to try the icy
upstairs bedrooms. They were in the parlour, curled around the
wood heater, swapping yarns. The room was dark except for the
glow from the stove. When the fellows looked up and saw Stan
ghosting in the doorway it gave them a start. He was shrouded
in frost from head to toe except he didn't have a head because
of the blanket. Once they recovered their wits and realized he
wasn't Becher's ghost or the Abominable Snowman, they made
room for him around the stove.

In the morning they took some gasoline and hot water to the
stiff, stalled truck, put the gas in the tank, the hot water in the
radiator. Stan soaked a rag in diesel, put it on the end of a shovel,
lit it, then held it under the transmission and rear end to warm
them up. One of the trucks gave him a tow and everyone was on
their way. But Stan's miseries weren't over. The hot water froze
before it got to his heater, and he had to drive with the window
open to keep the windshield from icing. There was no way to
keep himself from icing. When he got to Williams Lake it was
-40°F. He put the truck in a garage and himself in a hotel and he
stayed there until he thawed out. It was two days before he felt
warm again.

Every hill and gulch had a horror story. Betty (Graham) Linder
and her mother were on their way home one chilly night when
their car quit on Bear Head hill. They had some empty preserving
jars with them, and Betty filled them with hot water from the

radiator and the two women put them under their fur coats and managed to keep warm. They were rescued in the morning but Betty wasn't well for several days after. She said there was a cold spot in her stomach that wouldn't get warm.

Bill Graham had the closest call of all. The weather took a sudden mean turn in November 1954, and the temperature plummeted. Bill was driving a diesel-fuelled truck, and he stopped to snooze on the way home. The motor stopped and Bill didn't wake up until he was very, very cold. He couldn't get the truck going, but he wasn't far from the Redstone emergency cabin, so he walked to it. He managed to get the fire started but he was in poor shape. In the meantime, the family phoned around and learned he'd gone by Stuart's but not Chilanko. Olaf Satre, a Tatlayoko rancher who happened to be at Graham's, and Fred Linder went looking for him. They found him just in time: his feet were frozen so hard they sounded like sheep feet clicking across the floor of the cabin. He spent a few days in hospital, and he lost some toenails.

Chilcotin had its first contact with the Roman Catholic Church in the days of the fur trade. Oblate fathers visited Chilcotin reserves regularly from the 1880s on, and there were great gatherings at "Priest-time." In the 1940s, Missionary Sisters of Christ the King from Quebec set up shop at Anaham Reserve. They were strong, lively women. One of them, nurse/teacher Sister St. Paul, was a true pioneer. Seldom daunted, she tended to charge in where angels feared to tread. She became one of Chilcotin's legends but she almost became a Chilcotin tragedy. In January 1949, she and three other nuns were heading home from town in a geriatric Model A when they foundered in a blizzard on Becher's Prairie. When they didn't get home, Indian Agent Bill Christie was notified. He and his assistant, policeman Bob Turnbull, set out from Alexis Creek. Knowing that someone would look for them, the nuns wisely stayed with the car but unwisely ran the motor to keep warm. Turnbull was late, he got drifted in and was rescued by Bert Roberts who was ploughing the road. The nuns were unconscious when the Indian Affairs officials reached them. The men pulled them from the car and propped them against the snowbank. Sister St. Paul came to just after Turnbull arrived. "Well," she said, "it wasn't St. Christopher who saved us, it was St. Christie."

When the road across Becher's Prairie disappeared under the drifts, travellers struck out wherever, hoping they wouldn't be ambushed by the rocks that hid under the snow waiting to snag their axles. By spring there were roads all over the prairie, each studded with shovels abandoned by people who dug themselves loose and didn't dare stop once they got going. Bill and Irene Bliss once dug through a drift with a newly purchased dustpan, the only tool they had with them. Public Works put snowfence along the worst places. It was made of thin slats wired together and when the snow drifted against it, it looked like a tremendously long sea serpent undulating through a snowy sea. The fence led a double life. In the summertime everyone who could get their hands on it used it to keep porcupines out of gardens and children inside yards. Nothing could climb it, it was too wobbly.

# CHAPTER 13

# Rest and Refreshment

*We didn't expect anything and we didn't get anything. It was such
a vast country–who could do anything about that?*
—Anahim Lake merchant Isaac Sing

Distances shrank as the road improved, but until the
1960s it was a chore getting up Sheep Creek Hill and travellers
were ready for a coffee when they got to Riske Creek. After Fred
Becher died, his widow rented the place to Bert Roberts. In 1942
she sold it to Riske Creek rancher Geneva Martin and went to
England. Mrs. Martin renovated the place but the glory days were
over, and when Public Works rerouted the road over the hill,
bypassing Becher House, she gave up. The building sat for years,
genteely decaying, a monument to Chilcotin's unfulfilled dreams
of greatness.

Roberts, who worked for PWD off and on, built a store and
guest cabins a few miles up the road. In the late 1930s, Vancouver
businessman George Christensen (he started the Piggly Wiggly
grocery chain) embarked on an ambitious project to have a string
of resorts along the Chilcotin Road. He built a luxurious ten-bed-
room lodge at Riske Creek on a hill overlooking the plains. The
lodge cost $38,000 and there were four cabins besides. When the
war broke out Christensen went to Ottawa to serve (for a dollar
a year) on the Wartime Prices and Trade Board, and the Chilcotin
plans were abandoned. In 1941 he offered free use of the lodge
as a rehabilitation facility for servicemen in trouble for alcohol-
related offences, but the project didn't get off the ground. Tom
Rafferty bought the place and operated the lodge until 1965.

A road change left the TH Guest Ranch off the beaten track,
but that didn't deter guests or local overnight guests. Hodgsons'
drivers stayed there when they hauled freight into the Whitewater

country. Wilfred's fiancé Dru and his sister Betty Jean made one trip with him. After dinner Rene Hance escorted the women down a narrow path and over a little bridge to one of the rustic cabins, lighting the way with a flashlight. When they reached the cabin he opened the door and shone the light around. "Here you are," he said, "that's Billie Woods in the bed." He and the flashlight left, leaving Dru and Betty Jean in the dark. There was nothing to do but get into the already occupied double bed. In the morning their bedmate was gone. They learned later it was Hance's cook, Billie Woods, the wife of prospector Billy Woods. Dru never did meet her formally.

The Hance boys sold the ranch in 1965, almost ninety years after Tom pre-empted the property. The post office was in the family for seventy-six years. Rene was postmaster for thirty years, marriage commissioner and coroner for forty years, and a provincial court judge for eight years. The TH is still a guest ranch, and Nellie Hance's yellow roses still bloom there.

Most travellers had had enough of the road when they reached Alexis Creek, and they stayed at Tom Lee's or Jakel's. Lee served customers until he was well on in years, and his son Tommy was involved in the business early on, trucking freight and minding the store. Tommy liked to tell jokes (some of them awful), and he played the accordion, often providing the music for community dances. He had a unique style that didn't always adapt itself to playing with other musicians. His wife Louise went to Alexis Creek to work in the hospital, and after their marriage she helped mind the lodge and the post office.

Gus and Doris Jakel operated the hotel at Alexis Creek for ten years. Alexis Creek was seventy-five thirsty miles from the nearest liquor outlet in Williams Lake, so Gus, a leprechaun of a man with big ideas, thought a beer parlour might be profitable. Because provincial regulations required hotels to have ten bedrooms to qualify for a liquor licence, Gus added a second storey to Marshall's stopping house. He designed and installed Chilcotin's first water system, piping water from a spring-fed creek that falls from the hills above the village. The small beer parlour (15 by 24 feet) featured bottled beer, no glasses or beer on tap. Customers helped themselves if Gus wasn't around. The hotel plumbing included flush toilets, another Chilcotin first. The beer parlour had two doors, properly marked Ladies and Men, but there was only one room behind them. It had a proper convenience for the ladies, and a trough for the men.

Local travellers called the dining room the Apple Pie Cafe because Doris baked every day. Another of her gourmet specials was bologna sandwiches, a big hit at social functions. Being store-bought, bologna was considered a luxury instead of eggs, fish or game. The service at the hotel was casual. When Stan Dowling and two women passengers arrived at 4:00 a.m. after leaving Towdystan at 2:00 p.m. the day before, Doris got up to let them in. "You know where the sheets are, Stan," she said. "Show the girls so they can make their beds." He did and they did. Jakels charged $4 for a bed and a huge breakfast—platters full of bacon and eggs and pancakes, as much as anyone could eat. Gus tried everything to make the establishment profitable. An advertisement in the Cariboo Digest read: CHILCOTIN HOTEL. GUS JAKEL. HUNTING, FISHING, RIDING. REGISTERED GAME GUIDE. SUNSET TRAIL.

One of Gus's better known stories involved cattle. He "dined out" on that story, which involved half his cattle herd being killed by bears. The story was absolutely true: he once had a milk cow and calf, and a bear killed the calf.

After World War Two, Gus sold the store to Eddie Pigeon, a handsome, charming little man, the son of Cariboo pioneers. Eddie operated the store for the next twenty-three years. The building had a top storey with an outside staircase that served as the Alexis Creek community hall until the 1960s. Eddie held weekly movies in the hall for years, and dances and weddings and church services were held there. There were no kitchen facilities, so whoever was hosting the "do" would make coffee in a big kettle and get two strong men to carry it to the hall and put it on the wood heater. Liquor wasn't allowed in the hall so people kept their booze in their vehicles and slipped out now and then to have a nip. The more dedicated drinkers sat out with their bottles all night. There were no washrooms in the hall, and no outhouse. Men peed wherever. Women didn't dare squat behind a parked car or pickup for fear of being caught. They had a choice of

*Ploughing Becher's Prairie, 1930s. Bud Barlowe was the first power grader operator in the Williams Lake district. His job included snowploughing Becher's Prairie with a V-plough attached to the front of the grader. The machine was not really heated, and long days were the rule rather than the exception. There was no backup, so Bud usually worked on his own.*

*Road warrior. When Stan Dowling started his freight run to Anahim Lake, the road was full of mishaps. Once, in the truck pictured here, he got going too fast on Sheep Creek Hill. The truck fell off the edge, rolled completely over and ended right side up on an old road down below. The truck was damaged, but Stan and his passengers weren't.*

*Cookie's truck. Whenever possible, each Hodgson driver got his own truck. This 1947 GMC was Cookie Hodgson's pride and joy. He got mad if anyone else drove it. As well as suffering any number of flat tires each trip, trucks often broke down. Most drivers carried enough spare parts to rebuild the entire vehicle.*

*Upset. Gordon Jakel operated the garage at Alexis Creek but he also had a trucking business. Like all truckers he had his ups and downs. He rarely made a trip without running into some kind of trouble, even if it was only a flat tire. In this case he suffered a major upset.*

*Detouring the bull's nose. When the federal and United States governments got together to rebuild the road from Williams Lake to the Puntzi air base in the early 1950s, traffic had to detour in Bull Canyon while the crew blasted the rock bluff to widen the road by the river. The grade on the detour road was so steep the contractor, Northern Construction, kept a bulldozer on standby to pull trucks over it. The blasting crew used a bit too much powder to blow the bluff (they called it the bull's nose) and rocks landed in fields on the other side of the river. There is a provincial campsite near the widened spot now.*

*Tight navigation. Faced with the problem of hauling extra-long poles up Sheep Creek Hill in the 1950s, Wilfred and Cookie Hodgson used one of their dad's tricks. They put the poles on two trucks to get around the switchbacks.*

hiking to someone's house or down the road to the beer parlour. Pigeon's house was the closest. It didn't have inside plumbing, so Suse Pigeon's friends used her commode. It got pretty full on dance nights.

Gordon Jakel took over the garage after the war, and Irene and Sam Barrowman, retired professional entertainers from Vancouver, bought the hotel. Irene was a striking woman who wore her prematurely grey hair close-cropped. She always dressed in black and white. She had a strong torchy voice and could really belt out a tune. Sam, a portly little party with silver hair, accompanied her on the piano. They kept their hand in by performing on cruise ships. The Barrowmans were generous with their talents and willingly performed at community functions, but Sam was a tad tight when it came to business. For years he charged 15 cents more for a beer than anywhere else, blaming the high cost of freight. Chilcotin drinkers were ready to string him up when the Liquor Control Board caught up with him and made him drop the price. It seemed he didn't pay any freight, the breweries paid it. Sam had a wall between the washrooms in the beer parlour but he put "Pointers" on one door and "Setters" on the other. The LCB inspectors didn't like that either.

Until BC Hydro arrived in Alexis Creek in 1967, most places had a "light plant." The engines powering the generators came in a wide variety of models, and each one sang a different tune. There was the t'chug t'chug of the one lungers (one piston), the staccato yap of the little gas plants, the booming bass of the big diesels. Each plant was housed in an outbuilding and each was turned off at night by the last person to bed in each household. Everyone in the village knew who was up and who wasn't by listening to the engines. The Barrowmans lived in a little log house behind the hotel. Sam shut his plant off on his way home each night after closing the pub, leaving his hotel guests in the dark. On Sundays the pub was closed and he went to bed earlier, leaving them in the dark longer. Barrowmans sold the hotel to Pat Scallon and Joyce Russell in the late 1950s. The Scallons were Big Creek pioneers, and Mrs. Russell went to Chilcotin to manage the Chilco Ranch store. She loved to party. The social life at Alexis Creek was particularly gay while she was there. She organized cabarets (legal by then), card parties and picnics, and catered most of the events, preparing fabulous meals.

May Stewart's Log Cabin Cafe was in a hollow between the C1 Ranch house and Jakel's garage. May arrived in Chilcotin from England in the late 1920s to visit the Blisses. She was well educated and had served as governess to a highly placed English diplomat in Spain. She worked for the Ralph Ross family at Redstone, and when Ross died he left her enough money to start the Log Cabin. May was a hummingbird of a woman. She never walked, she darted. Her meals were excellent but the service was erratic. She couldn't do enough for people she liked, she was gracious with strangers, but no one else knew what to expect. Her living quarters were separated from the cafe by a curtain, and occasionally when she popped out to see who had arrived, she didn't like what she saw, and flapped back into the kitchen, never to be seen again. She took an active part in community affairs in more than one sense of the word. She could be counted on to volunteer whenever needed, but if something or someone displeased her, she was apt to take direct action. She gave one fellow a good drubbing with her shoe (he was almost a foot and a half taller and weighed twice as much as she did) and she went after another chap with a two-by-four. Both men were badly shaken by the incidents and neither ever knew exactly what he'd done to deserve it. The Log Cabin burned in the mid-1960s, taking May's life savings in cash with it. She got along as best she could after that, ending her days in the seniors' lodge in Williams Lake.

Alexis Creek had one attraction that was hard to classify. The Chinese workers who were so much a part of early Chilcotin life disappeared one by one, except for Kin Nauie, who worked for Alex and Anna Graham. When they sold the C1 Ranch to their daughter Frances and her husband Duke Martin in 1927, Kin Nauie, who had practically raised the Graham daughters, stayed with the place. Frances died and Duke remarried, but by then Kin Nauie was a ranch fixture. His claim to fame was his spectacular cussing. He swore at everything, but the milk cows were singled out for special attention. The C1 Ranch buildings were right downtown Alexis Creek, and Kin Nauie's voice carried well. "Bugger you sullakabitchee whores" was one of his milder epithets.

Strangers were astonished at the colour and range of his comments, which for years contributed to the local children's education.

Bill Bliss Jr. and his wife Irene (a Hance granddaughter) stayed on the original Bliss place, Willow Springs Ranch, which had cabins, a gas pump and a door that was always open to weary travellers. One of Chilcotin's excellent cooks, Irene always seemed to have fresh bread right out of the oven. She produced magnificent feasts, even for unexpected company.

Christina Stuart started minding the Redstone Store and post office when she finished school. She was a handsome woman but she rarely gave much thought to her appearance. She didn't let the weather dictate to her either. She rarely wore a hat or gloves and she ran back and forth from the store to the house in sub-zero temperatures with no coat, her bare feet in bedroom slippers. She was always civil in the post office, and kindness itself to neighbours, but she had mood swings with strangers. A couple on a fishing trip stopped for supplies one day, and when the woman asked for some bologna, Christina slapped the roll on the counter and began whacking slices off it with a big knife. One chunk fell on the floor. She picked it up, wiped it on the side of her skirt, and put it with rest of the slices. "You're not going to try to sell me that," the customer gasped. "I'm not going to try to sell you a damn thing and you can take your ass out of here," Christina snapped. She bundled up the bologna, put it away, lit a smoke, and stood glaring at the woman until she left. Tourists were attracted by the old vehicles and stuff lying around Stuart's, but if they stopped to nose around or take pictures, Christina pegged rocks at them.

The Grahams at Tatla Lake went into the hospitality business the same way they went into the store—in self-defence. Tatla Lake was not only at the end of the Chilcotin Road for years, it was also at the forks of roads into Eagle Lake, Chilko, Tatlayoko and Bluff Lake. Grahams had a bunkhouse for ranch hands and stray men, a cabin for the Native folk, and eight bedrooms in the main house. Guests ate with the family and the Graham table really did groan with food. Betty Graham, who married trucker Fred Linder, ran the store and post office. When Bill married, his wife Joy helped Mrs. Graham. The parlour was reserved for very special guests, the parson perhaps, or the Queen if she happened by. Joy and her sister Eve, both city girls, went to Tatlayoko in the 1930s to work for Ken Moore, who had a stopping place catering to miners. Eve married Tom Chignell.

There was a big gap on services between Tatla Lake and Anahim Lake. When the Dowlings moved to the McClinchy Ranch at Kleena Kleene, a man named Gordon McGinnis leased the store at Anahim. A short time later it burned down, leaving only Christensen's store at Clesspocket until a genial young Chinese Canadian named Isaac Sing moved to Anahim.

Ike was born in Merritt and his parents were naturalized Canadians, but BC laws restricted the activities of "Asiatics." Ike fished commercially at the coast until his licence was revoked (because he was Asian), then he operated a fish and chip shop in Vancouver. In October 1945, he went to Anahim Lake on a hunting trip with a friend from his fishing days, going up the mountain from Bella Coola. He met Lester Dorsey and his wife Mickey, and Thomas Squinas, and went back the next fall for another visit. When Dowling's store burned, the Dorseys and Squinas suggested Ike start a new one.

"There'll be a road to Bella Coola soon," Thomas said, and he convinced Ike he should build near the Indian reserve, where the road might go. He put up the logs for a store while Ike rounded up windows, doors and trade goods in Vancouver. Ike and his brother Jim set out for Anahim in mid-December. Harold Engebretson and his wife Alyce (Andy Holte's daughter) were living at the old Hudson's Bay post, and the Sings stayed with them until the store was livable. By the time the logs were up it was too cold to chink so they slapped mud from a nearby creek between them. The roof didn't leak because the snow on it melted and froze solid, sealing the cracks. When Hodgsons brought his trade goods, Ike dismantled all the packing boxes and used the wood and nails to hold the windows in. The door had no hinges, but it worked when the store opened in January.

It didn't take Ike long to learn the Chilcotin Two-Step. First his land application was rejected because he was Asian. Harold registered the property in his name and transferred it to Ike when the law was changed. Ike had hungry times. Nobody had much money anyway, and locals were used to shopping at Christensen's or having supplies trucked in. Then Andy Christensen bought the Dowling property, moved his store to town, and hired Alf Lagler to run it. That was stiff competition. For the first few years Ike lived in the store, with a sleeping bag for a bed and a camp stove to cook on. When he could afford to, he built log living quarters with a cellar underneath. He also got his own truck. He brought in apples, potatoes, carrots and onions which kept in the cellar until July. They were a special treat for Anahim residents who were unaccustomed to having fresh vegetables in winter. Ike introduced many firsts to Anahim, including the first refrigerator, a propane gas model. He brought ice cream home in insulated coolers. Revels were a big seller at 25 cents each. A supplier brought a truckload of soft drinks, another first, from Quesnel. It lasted all summer. After a long ride, cowboys grabbed a pop and a slice of bologna, and sagged to the floor to eat. In the winter Ike tried to keep the temperature in the store above freezing, which meant he was up every few hours stoking fires. The pop froze once in spite of his efforts, and the lids popped off. When the pop thawed, he put the lids back on with a bottle capper. If anyone noticed the pop was a bit flat, they didn't complain. Another first for Anahim was Ike's light plant, a gasoline-powered generator. It powered a water pump that provided for inside taps, a hot water tank and a gasoline pump—more firsts.

Ike built a bigger store, but he must have been going too quickly for the Drummer, because he got a good poke in the eye. The forest ranger ordered him to cook for a crew fighting a small fire in the area. When he got home he found some well-oiled cowboys wrestling around on the store floor. It made him mad. He threw them out and went to bed. One of the cowboys left a cigarette butt smouldering, and during the night the store burned to the ground. Ike moved back into the old building and started all over again.

CHAPTER 14

# Armed Forces Invasions

*The Dominion government rebuilt the Chilcotin Road from Williams Lake to Mile 130 (Puntzi Mountain) by fall. Completed fifty miles of new grade and alignment.*
–Public Works Annual Report, 1951

Governor Seymour led the first army to Chilcotin in 1864 when he came to quell the Tsilhqot'in rebellion. In the 1990s, the Department of National Defence (DND) and the Toosey Band are having differences over the military exercises that take place regularly on territory Toosey claims. In between, there have been three other army invasions, all of them more or less peaceful.

The first of the three came in 1942, when the telephone line between 150 Mile House and Bella Coola was replaced. When Japan attacked Pearl Harbor in December 1941, the Canadian government expected BC to be next, and Bella Coola seemed a likely target. The single telephone line dangling from the jackpines across Chilcotin was obviously inadequate for wartime emergencies so the Government Telegraph Services and the Department of National Defence built a new one. Army and air force personnel, many of the latter speaking only French, dug the post holes and put up the poles, but civilians were hired to haul poles and string wire, and worked as cooks, carpenters and road crews. The project was a bonanza for settlers who had had few chances to make a dollar during the Depression.

Much of the work was done in the winter of 1942–43. There were half a dozen different crews, about forty men in each camp. There were two army camps and one or two air force camps. Everyone lived in tents. DND and the telegraph people didn't know about the Drummer, so they were unprepared when the temperature dropped to -60°F in October and seldom reached zero from the bottom end for the rest of the winter. The lumber for the tent floors got rained on in Vancouver; it froze so hard the carpenters had the devil of a time nailing it. The temperature inside the sleeping tents was -28°F in spite of the heaters, and some of the crew nearly froze. One fellow had fourteen blankets and shivered all night. The crews shivered all day too, sawing and trimming the poles and planting them. The servicemen had to dynamite the post holes, the ground was too frozen to dig.

Every big truck in the country was hired to haul supplies or the treated cedar poles. Gordon Jakel went hauling poles when he came home on leave from the army and he was so busy he overstayed his leave. The local doctor signed a note saying he was sick so he wouldn't be listed AWOL. A few people lost their shirts on the project. The men who contracted to haul the poles up Sheep Creek ran into trouble, as did another outfit that tried to haul poles to Anahim Lake with two bulldozers and sleighs. Road crews were kept busy too, even though most of the hauling was done in winter.

Bella Coola residents were excited when the army built a road to the foot of the Precipice—they thought they had their road for sure. Pete Evyan from Chilcotin and Ernst Gyllenspitz were powdermen on the job. Gyllenspitz was a Swedish aristocrat who somehow ended up at Atnarko. He'd been a powderman for the CN railway and really knew his job. He said he could build a road across the Great Slide and the army engineer let him show how he would do it. With nothing but some powder and a wheelbarrow with Fred Engebretson to push it, Gyllenspitz made a narrow but perfect trail across the two worst sections. Tom Chignell, the West Chilcotin telephone lineman, witnessed the show. "I believed Gyllenspitz when he said he could build a road across the Great Slide," Chignell said. "It was wonderful what he could do with bit of powder." But the army wasn't interested in a road, just a telephone line, so nothing came of it.

Chignell had a special interest in the Precipice trail. He and a friend went down it once on motorcycles, the only wheeled vehicles that ever tried it. He didn't recommend it, as the two pushed and carried the bikes as much as they rode. Stan Dowling and Andy Holte hauled poles down the Precipice road, making a trip

a day from Anahim and back, and people used the route until it
sloughed in.

In spite of the cold and the misery, there were no serious
accidents during the line construction. But there were some close
calls; Alf Lagler had one of them. Dowling built a good-sized
garage at Anahim Lake, and when he came in late from a trip,
he'd run the truck in and Lagler would unload it in the morning.
Stan had just gotten up the morning after a trip when he heard
a whomp from the direction of the store. He looked out and saw
the stovepipe flying through the air and Alf flying out the store
door. Stan dashed over. Alf was coughing and choking. He said
the stove had blown up, and indeed it had. The jumbo heater
made of quarter-inch steel plate was in a thousand pieces. Bits of
plate had blown through the windows, the stovepipes were
smashed flat and the store was full of ashes. "What in hell happened?" Stan was amazed at the damage. Lagler didn't know. He
had lit the fire and it blew up.

It took a while to solve the mystery. Stan's freight included
blasting powder and caps for the post hole crew, and someone
came in the night to get them. The caps were in boxes wrapped
in newspapers, and whoever picked the stuff up chucked the
paper on the ground, missing one of the caps. Alf lit the stove
with the paper. He always leaned on the heater while it warmed
up and he'd just walked away when it blew. The camp foreman
asked Stan to keep the whole episode quiet or there would be
hell to pay, and he gave him a new heater made out of a gas
drum.

The new telephone line was a big improvement over the old
one, there were no more relays, no more howlers. The repeater
stations at the 150, Kleena Kleene and Bella Coola were staffed
by RCAF personnel during the war. People in each community
could ring each other but they had to go through the operator
to get "out." After the war Gwen Colwell ran the switchboard at
Kleena Kleene and Nellie Kinkead, later assisted by Vera Hance
(Rene's wife) and June Bliss, minded the switchboard at Alexis
Creek. The operators were supposed to work regular hours, but
they were on call for emergencies and their idea of emergencies
was elastic.

"If someone needs to phone out to catch the stage to bring out a bottle, it could be an emergency," Gwen explained. The operators took messages, sent messages, called around to locate people, and in general kept Chilcotin going. The new line shared one fault—or bonus—with the old line. People could still rubber, and the strength of the signal diminished in direct proportion to the number of people listening in. Rubbering was a favourite pastime, it kept people in touch. Mrs. Christie at Alexis Creek listened in so often people used to pick up the phone and ask her to hang up before they even said hello. Nobody said anything on the phone they didn't want the entire Chilcotin to know about.

Chilcotin was invaded for the second time in 1945 when the Polar Bear Expedition arrived for winter manoeuvres. Over a thousand men were involved in Polar Bear, which simulated an invasion from Bella Coola. Every kind of snow machine was tested—tractors, weasels, snow jeeps, a machine with tracks and skis, snowmobiles like small tanks, two-man motorized toboggans, all kinds of trucks, and about 120 horses. The manoeuvres began February 1. Convinced the equipment would wreck his road, O. R. Roberts, the district highways engineer, persuaded PWD Minister Herbert Anscombe to have DND sign a contract to have the road "returned to its original state" after the manoeuvres. Ordinance corps, signals, engineers, ski troops, pack horse troops, artillery, infantry and the air force participated in Polar Bear, while the Canadian navy and British and US services sent observers. The advance people arrived in Williams Lake in January and set up camp in tents. Staff Sergeant Alex Fraser, who was later to become Cariboo MLA and highways minister, was stationed in Williams Lake looking after stores for the army. One of his tasks was to have hay and oats for the horses flown into Anahim and Bella Coola.

The army trooped west on February 15, camping along the way and causing considerable excitement at Riske Creek, Alexis Creek, Tatla Lake and Kleena Kleene. The engineers did some road work as they went along, "affecting many improvements with tractors and bulldozers in spite of the adverse weather," according to the records. They put heavy shale on the road near Alexis Creek and built culverts. The last camp was at Anahim

*All dressed up. Chilcotin residents didn't let mud, mosquitoes or distances impede their social life. Nor did they let circumstances dictate their dress. Judging from their outfits, Rene and Vera Hance might have been on their way to concert when this photograph was taken in the 1930s. Actually, they were going to a football game. Vera said they got dressed up whenever there was an excuse. That's Gus Jakel behind the wheel.*

*Chilanko Forks, 1930s. Mr. and Mrs. Ole Nygaard, members of the Norwegian colony in Bella Coola, established the first ranch at Chilanko around the turn of the century. In 1912 Arthur (pictured here) and Anna Knoll bought the place sight unseen and made a large addition. Ducks Unlimited have the ranch now, and much of it is flooded for a waterfowl reserve.*

*Rebuilding the line. Until 1942 the telephone line linking Bella Coola and Chilcotin to the rest of the world was a feeble single line that sometimes dangled from trees. Fearing a Japanese attack from the coast, the Canadian government sent military personnel to install a better line. The project was a bonanza for local residents, providing better telephone service and a boost to the local economy. One of the big challenges was hauling telephone poles to the site. A contractor named Docherty, from Quesnel, is shown here with a truck full of poles in front of the Graham house at Tatla Lake. In winter, a bulldozer was brought in to assist.*

Lake right behind Dowling's, and the main air strip was built nearby. The soldiers were put on rations, and Dowling did a good business in chocolate bars and biscuits until his store was put off limits.

The Drummer saved some extra cold weather for Polar Bear. The snow on the mountain was fifteen feet and deeper. The exercise called for the enemy to "blow up" Canoe Crossing while the good guys fended them off. Thomas Squinas was hired to guide one batch over the Summer Trail. The expedition was filmed so he wore a uniform. The armoured snowmobiles sank in the deep snow, the jeeps and weasels foundered, so the troops skied to Bella Coola. Supplies, including wagons, were parachuted in to them. A packtrain of 117 horses carried essentials, some of them, such as 75-mm mountain howitzers, on travois. The enemy ski troops went down the Precipice assisted by Alfred Bryant and Bill Lehman (Jane Bryant's husband) who packed supplies for the officers. When the battle was over, navy ships evacuated the equipment and the men returned to Anahim via Atnarko and the Great Slide.

As Roberts expected, the Polar Bear equipment destroyed the road between Sheep Creek and the top of Norman Lee's hill. He wanted $20,000 to fix it. DND demanded a court of inquiry, but Roberts had before and after photos. The army paid up, but by the time the red tape was cleared away it was too late in the year to repair all the damage. PWD took the money Roberts couldn't spend and used it for other projects. Parts of the road through Riske Creek and Hance's Timber were a mess until the military came along the next time and fixed it.

The third armed invasion lasted fifteen years, and Chilcotin literally stood on guard for Canada and the USA. During the 1950s the American and Canadian governments expected the USSR to launch an atomic attack at any minute. In response to the threat, they set up a series of early warning air bases to alert the military in time to launch a counterattack. A site near Puntzi, which had figured so largely in the war of 1864, was chosen as one of the thirty-four stations in the Pine Tree Defence line. The Puntzi Mountain air base (everyone called it Puntzi) came into being in 1951 and the military rebuilt the Chilcotin Road between the base and Williams Lake. The contractor, Northern Construction, hired local men and machinery. Bill Bliss, Harold Stuart, Gordon Jakel and dozens of others worked on the project.

The base was built on forty acres of the Knoll ranch at Chilanko Forks. Arthur Knoll was living there, but his son Ollie (Alvis) actually owned the place and he had his headquarters across country at Chezacut. The base was well established before Ollie and lawyer Jack Cade rode over from Chezacut to investigate. Security was tight at the base. When civilian worker Phil Robertson delivered the mail inside the base, he was accompanied by an armed guard. The officers were taken aback when Knoll and Cade rode in from the bush. They immediately took steps to secure the area. Arthur Knoll, who did not mellow with age, took exception when they locked a particular gate. He put a lock on it too. Unable to stop him from doing it, base officials finally gave in and left the gate unlocked. There is no evidence that the unlocked gate threatened national security.

Old Knoll kept pigs, and the cooks at the base saved scraps for them. They felt sorry for him, thinking he was poor (he wasn't) and neglected (he fought with all his family), and they slipped him goodies too. He did well from the base, but it was years before Ollie was paid for the property.

In terms of population and services, Puntzi was the biggest and most modern centre in Chilcotin. It was seven miles from the barracks to the operations buildings and a bus made six trips a day taking crews back and forth on shift changes. The base had a 6000-foot air strip, the second longest in the province, and there were thirteen D8 bulldozers on hand to clear snow off it in winter. Radar with a 150-mile range scanned the skies all day every day and the base was in close contact with the air defence base in Seattle as well as the Royal Canadian Air Force base at Comox. Along with the administration buildings and four radar bubbles, there were living quarters on the base for the 100 US airmen. The base hired local people too and the community of Puntziville sprang up to house the civilians, many of them Chilcotin born, such as Robertson and Clarence Mackill. Harold Stuart did a lot of work at the base and so did others. There was a school at Puntziville and for a time one of the US officers' wives taught there. Settlers drove miles to attend movies, parties, the pub and ball games at Puntzi, and the base was a bonanza for Chilcotin and Williams Lake. Trucks ran steadily between the laketown and Puntzi. Hodgsons couldn't begin to keep up with the freight.

The US eventually turned the base over to the Canadian government, and it wound down, closing completely in 1965. Most Puntziville residents left, but others came later. Some things from the base were sold, some found their way to local use, but much of it was torn down and destroyed, much to the disgust of residents. The Drummer is still trying to reclaim the land where the base was.

# CHAPTER 15

# The Mechanics

*If you have a fast way of fixing something people are happy. It's not what you know, but what you do. It's amazing what you can do with a chunk of haywire and a prayer.*
—Harold Stuart, Redstone

Motor vehicles kept nosing their way west, although they didn't get much encouragement. For years the only service station in Chilcotin was Jakel's. The Drummer called Gordon Jakel home after the war, and he took over the garage. He sold Shell Products. He had underground fuel tanks and installed Alexis Creek's first power plant which powered Chilcotin's first electric gas pump. He lived over the garage, which was handy for motorists who didn't mind rousting him out in the middle of the night if they wanted something. Gordon was a good-natured man, but he got tired of that, so he rigged the pump so people could help themselves. He trusted them to pay the next time they came by, which they did. Gordon usually had somebody hired to help and he had a freight truck too. Many young fellows got their start working for Gordon. The original log building burned in 1970, the same year Gordon received a silver plaque from Shell Oil for thirty years' service. He built a modern garage on the site. When he retired, he stayed at Alexis Creek.

Jakels went into the garage business on purpose but they were the exception. Other service stations just happened. The Harvey place was a few miles west of Alexis Creek, near Telford's place. Tom Harvey Sr. was Anna Graham's nephew. His son, Tommy, didn't actually have a garage, but he was a whiz with a lathe and he did machine work for people.

Then there was Redstone. For over seventy years Redstone was the mailing address for the Bayliffs, Blisses, everyone else within twenty miles, but as far as the dot on the map was concerned,

Redstone was Stuart's. That's what it should have been called anyway since the outcropping of red rock it is named for is twelve miles up the road. Except for some years going to school in Vancouver, Andy and Hettie Stuart's son Harold, born in 1921, and daughter Christina, two years younger, spent their lives at Redstone.

Harold was born to the Drummer's beat. He operated his garage for over forty years and he was outstanding on two counts. First, he had few peers as a mechanic. He could fix anything. If he didn't have the right tool, or the right part, he'd make it. He was expert at "mickey-mousing"—making emergency repairs to get the motorist home. His emergency repairs often outlasted the vehicle. Second, Harold was stubborn. In a country of pig-headed people he had no peers. He would do anything to help a neighbour, but he couldn't be rushed, pushed or coaxed to do anything he didn't want to do, and everything was done his way, in his own sweet time. If customers didn't like it they could stuff it, and he let them know where.

Andy bought his first truck in 1923, and he had two when Harold quit school in 1939. Harold began driving one of them and found it was easier than anything else he could do to earn a living. He liked monkey wrenching so he started doing his own repairs. Harold operated on Chilcotin time. "If something buggers, it only takes time to fix it. Nothing is lost if you can't fix it, only time. If you can fix it, you've saved money," he figured. Word got around that he was handy and he began fixing things for neighbours and for people whose vehicles broke down on the road.

After Harold started mechanicking, traffic had to elbow its way through the Stuart establishment, dodging bits and pieces of equipment and vehicles of various vintages crowding the right-of-way on both sides of the road. For over forty years assorted government officials tried to get Harold to clear the roadway but he never did; it just kept getting narrower. Eventually they moved the road.

Harold didn't have a proper garage building, just a little log cabin crammed with tools. His parts department was outside and he had an eclectic collection of artifacts piled all over the place.

It looked like a junkyard run amok, but he had everything he needed in that pile and he could find it if he searched long enough. He had a gasoline pump too, though he rarely sold gas. There was a sign on the pump saying "Honk for Service" but if anyone honked it made him mad and he wouldn't serve them.

Harold was self-taught and thought of himself as a diagnostician. "I can always find out what's wrong with something but don't hurry me, I have to play around with it until I find out how it works," he explained. He said there were two or three things he knew he couldn't do. No one ever found out what they were. He delighted in being rude to people he didn't like, and he didn't like big shots. "They want things done on Sunday when they know they can't get it done anywhere else. Americans are bad, sometimes if they're really in trouble I help them, but I give them hell. The worst are the snots who don't know me on the street in town, they walk right by without saying boo, then come crawling when they're stuck. They can stay stuck."

Harold did the big work outside. Even the most macho old-timers marvelled at his ability to do what was often delicate work in sub-zero weather, bare-handed, lying or kneeling in the snow for hours on end. He was a night person to boot, and working until the wee hours in the morning when it was coldest. Rain and mud didn't faze him either. He bought a big steel building from the Puntzi air base when it closed, and he took it apart and got it home, but he never got around to putting it up.

Harold's idea of money was elastic. He charged what he felt like, and rates varied from friend to foe. Sometimes he would do an incredible amount of work for little money, sometimes he would charge an incredible amount of money for little work. He didn't charge his Tsilhqot'in customers at all if they were broke but if he caught them faking, he charged them double. He soaked big shots. Once an American tourist wanted a certain bolt. Harold rooted around in the parts pile and found one. He charged 25 cents for the bolt and $5 for hunting for it. In later years the store was broken into so often Harold put heavy wire on the windows, but few thieves had enough courage to tamper with the parts pile.

Harold's sister Christina was a fair mechanic too, and when it came to independence, they were a matched pair. After Harold married, he and his wife Marcella and their eight children lived down the road from the business. Christina lived in the big house, across the road from the store.

One hot August day, Ed Wallace, the Anglican minister, was returning to Williams Lake after a week visiting his west Chilcotin parishioners. He left Tatla Lake at 10:00 a.m. in plenty of time to get home for supper. It was a sunshiny day. A few cauliflower clouds watched his dusty progress and all was well until a tire went flat about ten miles short of Redstone. His spare was a sorry thing, unlikely to get him home, so he stopped at Stuart's to get the flat fixed.

There was no sign of life at Redstone. A vagrant wind was sending dust devils down the road and making strange frying noises as it rose and fell in the cottonwoods by the river. Wallace walked across the road to the store. Inside it was as cluttered as Harold's shop, and it was dark after the bright sunlight. Wallace heard thumping noises, and found Christina sorting mail in the post office cubbyhole at the back of the store.

"He's at the shop," she said without looking up.

Wallace tiptoed out, blinking in the bright sunlight, and went back across the road.

"Hallo there," he ventured.

"Hallo there," replied a muffled voice. Harold emerged from under a derelict truck.

"It's a beautiful day," the minister said. He knew better than to jump right into business.

Harold agreed. Conversation came to a halt while he fished a battered package of cigarettes out of his shirt pocket, shook one out, put it in his mouth, and exchanged the package for a long wooden match which he lit with his thumbnail.

He was a gnome of a man, husky, dark-haired, unkempt. In later years he grew a big, bushy, slightly evil-looking beard. This day his blue jeans were black and shiny and grey woollen underwear peered over the collar of his shirt. (There were two kinds of Chilcotin men, those who wore long johns, and those who did not. Those who did rarely shed them on the theory that what keeps heat in keeps heat out. Harold was of this persuasion.)

With his smoke lit, he was ready to chat. Wallace told him all the upcountry news he could think of, then edged over to his car and opened the trunk.

"Troubles?" Harold asked.

Wallace showed him the tire. Harold inspected it, agreed it was "bad sick" and took it over to the shop. He was just starting to work on it when a dusty pickup truck rolled in, so he stopped to visit with the newcomer. As the cigarette ritual was repeated, Christina strolled across from the store and plumped herself into the conversation. She, too, lit up a smoke. They discussed the weather (great for the haymakers), foreign cars (we won the war but the Volkswagen is Germany's revenge), and the state of the road (the grader is at Chezacut). The latter topic gave the newcomer a chance to state his problem.

"Those damn rocks on Ross's Hill got my oil pan," he said.

When his smoke was done, Harold scrunched himself under the pickup to inspect the damage. Then he went into the shop, poked around, found what he wanted, crawled back under the truck, and began pounding at the oil pan. Christina toed her cigarette into the dust, went over to the shop, and began patching Wallace's tire. Work was proceeding nicely on all fronts when a blue delivery sedan screeched in and a young woman leaped out.

"I didn't think I'd make it here," she yelled.

Harold popped out from under the pickup and Christina dropped the tire. As the dust settled it was obvious the car's engine was steaming.

There was more conversation; more cigarettes were smoked. When the car cooled down, Harold lifted the hood. Muttering about "a sick hose" he headed for the parts pile. Christina finished her smoke and crawled under the pickup. She was dark and chunky like her brother and often as untidy. This day she was wearing a dark green sweater and a brown skirt that might once have been pleated. Her dusty bare legs stuck out from under the pickup as she banged on the oil pan. Harold's mining expedition was successful. He returned with a piece of hose and went to work on the car.

Time was sliding by and Wallace still had a three-hour drive ahead of him. At 2:00 p.m. he decided to patch his tire himself. The woman was conversing with Christina's legs but the pickup driver hovered over Wallace offering advice and encouragement. By the time the tire was repaired and re-mounted, Christina and Harold had traded jobs. She was bum-up under the car hood and he was toes-up under the pickup.

"What do I owe you?" Wallace asked Harold's toes.

"Nothing," they replied.

As the minister left, a truck full of people pulled up to the store. Christina abandoned the car to tend to them and the two garage customers followed her, their feet making little puffs of dust as they crossed the road.

"I just came for the mail," Wallace heard the woman say.

"And I just stopped to get my tire fixed but I came away with another Stuart story," Wallace thought to himself.

Just as Redstone was Stuart's, Tatla Lake was Graham's although there was more to it. Bill Graham was another top mechanic. The school at Tatla Lake stopped at grade eight and that's where Bill's formal education ended except for some courses he took later. Even as a child he was gifted mechanically; one school inspector was so impressed he made a point of getting special books for him.

A stocky redhead like his dad, Bill was a quiet, friendly, sensitive boy. The young Grahams divided the Tatla Lake operation among themselves. Alex, the youngest, was the rancher; Bill was the mechanic; and Betty, the eldest, ran the store and post office and helped in the house. The garage was intended to keep the ranch equipment running, but neighbours kept bringing things for Bill to fix, and travellers broke down along the road, so he eventually started charging. He had a proper garage building, he was well organized, and he usually had at least one hired helper.

Grahams had a car and a truck by 1935; in 1938 they acquired a tractor from Ken Moore, a Tatlayoko rancher. It was the first tractor in the west Chilcotin and Moore couldn't get along with it. Bill made sweeps for it from angle iron and this and that, and they pulled hay with it for years. It was faster than anything else.

Bill wasn't a talker, but he spoke volumes with a "hmm." That "hmm" could mean anything—hello, goodbye, yes, no, go to hell— and it was always very clear from the tone exactly what he did mean. Like most Chilcotin men of his time, Bill was a drinking man. It had no effect on his mechanicking and it really miffed his peers that he could do better work when he'd had a few than they could do cold sober. There are many stories of how Bill fixed the unfixable or found the trouble when no one else could. He often overhauled equipment by stripping it down to the frame and rebuilding it, making new parts if he had to. He was working on a job near Alexis Creek toward the end of the war when the flywheel went on his D6. Unable to get a new one anywhere, he turned one out on his lathe. He made a template out of cardboard, copying an old gear, then he went to Seattle and bought a milling machine and made the gear. It lasted as long as the machine did.

Bill's first bulldozer was a D6. It was intended for ranch work but he spent many hours ploughing snow and building road for Public Works, and that's another story. Bill also hauled the freight for the ranch and helped build the Freedom Road from Bella Coola, but that's another story too.

Bill was a shrewd businessman who didn't mind playing the rube if it served his purpose. When he was shopping for a new bulldozer, he took his cousins, Phil and Zander Robertson and Tommy Robertson, to Seattle. He had them check prices with all the dealers, then he made the rounds by himself. The dealers thought they had a live one—a rich hick—and he let them think it until they got to the bottom line. He ended up getting a very good deal on a D4.

Fred Engebretson at Towdystan was another talented mechanic. He wasn't really in business, he didn't have a sign out or anything like that, but local people got him to fix things. He helped people in a pinch, too, like the time in January 1937, when Stan Dowling got into difficulties with his new two-ton Ford. The truck was an oil gobbler, and Stan was coming home with a full load when it ran out of oil near Nimpo Lake and the motor seized. After some delay, the Ford dealer sent out a new motor, Hodgsons hauled it to Kleena Kleene, and Stan picked it up with the sleigh. In the meantime Fred rigged up a tripod beside the frozen lake and, working in crotch-deep snow beside a campfire in the nippy January air, he and Stan took the old motor out and put the new one in.

*Horse power. If there was no tractor or bulldozer handy, and there seldom was, horses came to the rescue. This photograph shows Gary Tyrell, John's son, being pulled out of the snow at Tatalayoko.*

*Road camps. Before the late 1950s, when Public Works began providing camper trailers for the crews, it was every man for himself when it came to accommodation. The crews had to move camp often, and the big trick was to find a handy water supply. There was no refrigeration so the workers relied heavily on canned food for grub. Butter melted, bread went mouldy and everyone got sick of bologna and Spam. When it rained the tents leaked on the bedding. Pictured here is Bill Graham's road camp between McClinchy Hill and Towdystan in 1947.*

*The Big House. Bob Graham cut his own lumber to build the "big house," which served as the Graham home and a stopping place known for its delicious meals.*

*Stuck at Riske Creek. One of the Chilcotin's most notorious bog holes was right in front of the Riske Creek store in springtime. On this occasion it victimized Hodgson driver John Tyrell, who needed a bulldozer to get back on track.*

*To the valley below. This was the dreaded Bella Coola Hill in the early 1950s. It is much wider now, and the surface is good gravel instead of loose rock, but it is just as steep and just as high. Some say the trick is to travel it at night when darkness covers the terrors.*

Fred never hurried. He said he had more time than anything else. One cold winter day he was doing something or another when a car stopped and the driver asked if he had any gasoline. Fred said yes, and went on with what he was doing. After some time elapsed, the traveller asked if he could buy some gasoline. Fred said yes. The traveller said could Fred tell him where the gas might be located. Fred said yes. The traveller asked where, and Fred told him. The man finally found the gas drum, and figured out he had to siphon the gas out of it himself. He was a bit leery about the setup.

"Are you sure there's no dirt or water in the drum?" he asked.

Fred finally looked up. He shook his head pityingly. "There might be dirt in it, and there might be ice, but there sure as hell is no water in it when it's this cold," he said.

Towdystan didn't look very progressive during Fred's reign. The original building (TE 1900 is carved on the lintel) squatted in the tall grass, sporting the sign "Fred's Roadhouse"—a gift from one of Fred's regular guests. The "new" house, weathered dark grey, and the elderly outbuildings sprawl comfortably around the clearing. Fred had the first mechanized ranch in west Chilcotin because he was allergic to horses. He got the first tractor he could, a 1928 Oliver on tracks. He used it for everything, travelling in it in winter and spring because it rarely got stuck. He cowboyed with it too. A few years later he bought a Massey-Ferguson wheel tractor from Dowling and he was still using both of them in 1989.

"I got ahold of some of their relatives for parts," he explained, adding he had to "manufacture a few pieces" himself.

Fred called himself a good "haywire" man and really did wonders with machines. However, he had some trouble learning to drive a car. Pan Phillips, who ranched in the remote Blackwater country, sold Fred his Model A Ford. Dowling tried to tow it home for him, but Fred steered it the way he steered his fish boat with a tiller. He got the car going every which way, and he mowed down so many trees Dowling loaded the car and Fred on the truck and hauled them both to Towdystan.

Earl McInroy had the first garage at Anahim Lake, but he didn't intend to. He went to Chilcotin in 1934 and worked here and there, first at Chezacut for Arthur Knoll, then in the Anahim

Lake country. He joined the services in 1939 and after the war he married Thelma Engebretson (Fred's sister). They lived in Harrison Hot Springs for a while, and their son Butch was born there, but in the fall of 1947 they moved back to Chilcotin and bought property at the west end of Nimpo Lake. Anahim ranchers were in the habit of parking their equipment after haying and forgetting about it, only to discover something was wrong when they went to use it the next year. Then they were in a hurry. They took things to Earl to fix, and as it was a long way between places, they stayed until the work was done. Sometimes several customers came at once, some with their families. If it took Earl a week to fix everything, they stayed a week. Thelma got tired of cooking and the free boarders took all Earl's profit. When Butch was six, they bought land "downtown" Anahim Lake and Earl built a garage and a couple of guest cabins.

He did everything the hard way at first, with simple tools and brute force, such as using hammers and crowbars to change tires, but he had a few conveniences. He built an air compressor using the motor from a gasoline-powered washing machine, and he provided a mobile service—he carried an acetylene torch on his saddle horse and travelled to ranches on emergency calls.

No story of the Chilcotin Road would be complete without a mention of Art Evans, the man who converted the entire Chilcotin to General Motors products. When Art arrived in Williams Lake shortly after World War Two, there were few if any GM products west of the Fraser. By 1951 there was nothing else. Art, who was born and raised on Vancouver Island, had an extensive background in automobile sales and service when he joined up with lawyer Jack Cade to open Williams Lake Motors. He was a hit from the start, although he had his work cut out for him in Chilcotin. GM wouldn't take orders without a substantial down payment, a new idea for ranchers who were used to jawbone. A slim, wiry man with an overdose of nervous energy, Art was called the Silver Fox—partly because of his silver-streaked black hair, partly because of his super salesmanship.

He knew he wasn't going to get rich selling cars. Parts were the cream, and the Chilcotin Road guaranteed a brisk business in parts. He honoured all claims, even questionable ones, and some of the old-timers had odd ideas. Charlie Moon had a Chev coupe, and years after he bought it, he broke down near Four Mile. He walked to Don Mackay's and phoned for someone to come fix his wheel. When Art got there he found the tire on the wheel in question was so thin he could poke his finger through it. The old man thought there should be a few more miles on it.

Art went to his customers. In 1951, George Powers sold his place at Charlotte Lake to Jim and Mary Ann Ross of Redstone, and he and Jessie moved to the old Knoll place at Chilanko. Someone thought Powers might want a vehicle now he was out on the main road, so Art jumped into a brand-new pickup and headed west to see him. He got to Chilanko after midnight. No one was stirring. The house was a big rambling affair and Art didn't know which door to knock on. After floundering around in the dark yard wondering what he was doing there, he picked a door and pounded on it. After a bit the old man came to the door in his nightgown, carrying a candle. He invited Art in, stoked up the stove, and in the shadowy kitchen, by the flickering candlelight, Art made his pitch. Powers knew everything there was to know about horses, but he was clearly out of his depth on motor vehicles. He listened politely. When Art ran out of steam, Powers asked what a new pickup cost. Art told him, expecting to dicker, but the old man got up, took the candle into another room, and came back with a cheque for the full amount. For once Art was speechless.

Art left about 4:00 a.m. He was down the road a ways when his conscience got him for selling a naive old man an expensive truck sight unseen. He drove back and got Powers out of bed again. It was still pitch dark. Art insisted Mrs. Powers look at the truck too. Still most courteous, Powers fetched his wife, found a flashlight, and with boots on and coats over their nightgowns, the two old people inspected the pickup. Powers walked all around it, patting the fenders and kicking the tires.

"It's got good rubbers," he said finally. Mrs. Powers didn't say anything.

The truck outlasted the old couple. It sits rusting quietly beside the old barn at the Gables, where Jessie spent her last days.

When Art delivered a new four-door sedan to Eddie Ross, he

took a bottle along to celebrate, as was his custom. Eddie and Peter Ross operated a ranch a few miles up the Chezacut Road from Redstone, established by their father Ralph, one of Chilcotin's earliest settlers. Both were married and had large families. Eddie, Pete and Art were on their second or third drink when Art realized the noise he was hearing came from several small Rosses who were attacking the new car with sticks and stones. Art was upset.

"Eddie, look what those kids are doing to your car," he sputtered.

Eddie got up, looked out the window, shrugged and sat down again. "Kids gotta play with something," he said.

That's when Art realized he never would understand the Chilcotin Drummer.

CHAPTER 16

# Winning the West

*If we're supposed to build a road, we'll build a road.*
—Louis Holtry, Anahim Lake, 1942

The road between the McClinchy Hill and Anahim Lake was a poor thing. Public Works didn't bother spending any money on it.

Stan Dowling was the only regular traveller on the road, and he coped with it. When he sold his trucking outfit to Hodgsons and moved to the McClinchy Ranch, Hodgsons coped with it too. James Mackill had a 1927 Dodge with high clearance and when road conditions were right, he took fishermen and hunters to Nimpo and Anahim, but there really wasn't much other traffic. Mackill died in 1942, following an operation in Vancouver. His son Clarence took over the lodge, and the road work, but nobody fussed about the west end. Little was done with it until Alfred Bryant took it upon himself to remedy the situation.

The Bryants' move to Anahim Lake from Tatla Lake did not do much to improve their situation. Phyllis Bryant took the three older children to Williams Lake for high school, supporting them by operating a shop and making a few extra dollars as a pianist. Jane went on to become a registered nurse and dedicated her life to caring for Anahim Lake area residents. For years she was the health nurse for the Southern Carriers at Anahim and Ulkatcho (and for everyone else), covering a 100-mile radius around Anahim on horseback. A legend in her time, she was posthumously awarded the prestigious Red Cross Florence Nightingale medal for her service as the frontier nurse.

Alfred had some cattle, and he guided, and he liked to stir things up once in a while. In 1939, he took on the Anahim Lake Stockman's Association. Most West Chilcotin landowners belonged to this body. It met sporadically and most members thought its main purpose was to keep newcomers out of the country. They objected when any stranger wanted range or land, even if nobody else was using it. The same people were elected to the executive year after year until Alfred introduced the novel idea of a secret ballot instead of a show of hands. This resulted in some new faces, one of them Alfred's. He was elected secretary and immediately began using his official position to agitate for a road. When writing letters to politicians and Public Works officials produced no results, he convinced the stockmen they should lobby for a road foreman at Anahim Lake. The stockmen didn't take this proposal too seriously. They didn't want the road improved. Nevertheless, Andy Christensen nominated Louis Holtry for the job. The others went along with it, more as a joke than anything else. Alfred sent the suggestion to Williams Lake and the stockmen were dumbfounded when PWD hired Holtry.

Sometimes the Chilcotin Drummer missed his cue, as he did with Holtry. A latecomer to the Anahim area, Holtry was not the Chilcotin type. He was a little man with a quick temper, and he had skills the others didn't appreciate. Anahim ranchers copied his method of stacking hay, but they didn't give him credit for it. They did not copy his method of whip-breaking horses. Nobody had seen it done before and they were horrified by it. All in all, Holtry wasn't popular in the community. What the stockmen didn't know was that he had built dirt roads back home in South Dakota. Instead of making a fool of himself as road foreman, he built roads.

The first thing he did as foreman was call a meeting of the stockmen and tell them his plan. "Every outfit will contribute one man to work on the road. They'll be paid the regular wage, but they'll have to work long hours, and Sundays too, so we can get the job done," he told the somewhat sour gathering. Everyone more or less agreed, even Lester Dorsey, who always did his best to be a thorn in the ass of any kind of progress. Like most Anahim men, Lester relied on the poor roads to keep people out of the country. Holtry didn't waste any time. The first thing he tackled was the road that went around Cahoose Flats and Three Circle. He changed it, following Dowling's road, and it made a wonderful difference. When it was finished the trip from Fish Trap

*Bayliff's and the road. Hugh Bayliff built his "big house" between a high bluff and a dampish meadow. He included a polo field nearby–he and his neighbours used to play but it never caught on with the rest of Chilcotin–but he did not give any thought to the location of the road, which curved around the house almost in a U. In Hugh's day there* were so few cars it didn't matter, but as traffic increased so did the problems. The dust was awful, especially from the big logging trucks. It hung in a cloud over the ranch and could not be kept out of the house. And motorists new to the road did not always know what to do when they whooped around a corner to find themselves in the middle of a herd of livestock. There were two bad accidents near the house, but nothing was done for years because there was no easy, inexpensive way to move the road.

*Pablo Creek, 1950s. Although it is on the east side of the Fraser River, Pablo Creek is officially on Highway 20, and it was one of the hazards faced by drivers heading west. The creek flooded frequently, washing out the road–which wasn't much to begin with.*

*The Bayliffs and neighbours. Left to right: May Stewart, an English governess who worked for Redstone rancher Ralph Ross and later operated a restaurant at Alexis Creek; Clifford Olsen; Gay Bayliff, the second generation of Bayliffs at Chilancoh Ranch; his youngest son Tony; Mrs. Kathleen Newton, who ran the neighbouring ranch for years after her husband died, then left the place to her nephew Tony; Dorothy Bayliff, Gay's wife; their son Tim, the third generation on the ranch; Jack Bliss, who managed the Newton ranch for many years. There is now a fourth generation Bayliff at Chilancoh–Tim's son Hugh.*

*Play up, play the game. Hugh Bayliff and his neighbours playing polo. Bayliff had his own polo field on the ranch, and his neighbour Reginald Newton raised Arabian horses for the game–which somehow didn't catch on in Chilcotin.*

*Passenger service. Two passengers–three if they were small–could ride in the cab of a truck. Any more had to climb in behind. John Tyrell (Teapot) had a "box" passenger on this trip. "It was bumpy and dusty back there in the summer, bumpy and cold in the winter," Wilf Hodgson recalled. "But we poked passengers in, tied down the tarp, and charged them for it."*

to Anahim took two hours in summer. It had taken up to two days.

Over the next few years, between breakup and freeze-up, with time out for haying, Holtry kept both Bill Graham's and Fred Brink's bulldozers busy. He kept two shifts working whenever he could. He got up at 4:00 a.m. to cook breakfast and he worked with both shifts. Sometimes in the afternoon he'd sit on a stump, nodding, completely pooped out. He ignored orders to stop working if he hadn't finished the job he was working on, and once the district engineer had to drive to Anahim and lay off the crew himself.

Holtry was frustrated by the lack of communications. "By the time I receive permission from you to fix the mud holes they have dried up," he told the district engineer in 1942. And he worried about choosing new routes. He never did get an engineer to come out to guide him. In 1947 he had Bill Graham rebuild the road from the top of McClinchy Hill to Fish Trap. Holtry ran a line following the telephone poles, skirting the mirey spots. The Cat worked ten to sixteen hours a day, seven days a week, from June 6 to July 28, with time out for breakdowns. Phil Robertson, a teenager at the time, ran one shift. Sometimes Holtry's wife Babe would visit camp, and Bill's wife and mother would drive up with care parcels of food. Phil would go fishing whenever he could, the fresh fish a treat after the usual fare of canned food, but in general, life on the road wasn't easy.

The crew—Graham, Phil and Holtry—camped in a large tent which they moved along with them. They were later joined by local ranchers Howard Paley and Bob Smith. They started work on June 7, starting at "Mac" Hill, and working west. Road crews were expected to be sociable whenever someone came along, and Phil wrote in his diary that there was plenty of action (in Chilcotin terms) the first week on the job. Howard Paley came up with Dowling to join the crew as the rock picker; Lester Dorsey and the Engebretsons returned from the Williams Lake Stampede; and the Brink brothers came by to borrow Graham's trailer to haul lumber, promising to return it before the road crew moved camp.

"It's turning cold tonight, also a clear sky. Will probably

freeze," Phil wrote on June 10. He was right. It did freeze. There was ice on the water pail in the morning and the crew dug out their winter underwear. On the first shift (Phil's) the cat worked the road; on the second shift (Graham's) it pulled the grader to smooth things out. It was awful going. The road from the top of McClinchy Hill is one of the rockiest in Chilcotin. When Graham and Holtry attached the pull grader behind the Cat to smooth out the work Phil had done earlier in the day, Phil noted they "bounced around quite a bit." That was one way to put it; Holtry had to hang on for dear life to even stay on the pull grader.

By June 11 the crew were "out of the rocks" to Caribou Flats. "The flats are about two miles long. It would make a nice hay meadow but I guess it is like the road, plenty of rocks," Phil wrote. A shower of rain brought out the mosquitoes, which "bit the hell" out of the road crew. There was no more traffic on the road to interrupt work until June 18 when "a whole string of Indians went past and one policeman" on their way to the Anahim Stampede. Laketown residents Joe Gillis and Bill Sharpe went by with the Hodgsons' truck to set up a hot dog stand. The next night it froze hard enough to freeze the milk, there were a few snowflakes in the afternoon, and it was cold all day. Two mounted policemen, the Dowling truck, the Brinks and some others went by, stampede bound. By then the crew had reached the divide. "Bill is going down the other side, he says we will have a downhill push on it now," Phil noted.

On June 20, the crew left the Cat all ready to go and headed for stampede, only to find there was nothing going on because someone had turned out the bucking horses. "We had supper at Harold and Alyce Engebretson's place, I think mostly everybody turned up there sooner or later," Phil reported. Everyone went to the dance, the music on this occasion provided by records on a loudspeaker. "Mostly everybody was tighter than heck last night, bottles everywhere a person goes." Phil, who survived life in Chilcotin seldom taking a drink, was in a better position than most to notice what was going on. He said the next night was a wild one, with everyone really whooping it up. West Chilcotin pride dictated that men drank as much as they could, and slept as little as they could at Anahim stampedes. It was considered fair to lean on a tree to snooze, and passing out didn't count. For many stampeders, the rodeo events were just an excuse for a party. "Mostly everybody was pickled to the eye brows. I stayed up all night, I left Anahim Lake at 7 o'clock and arrived at Towdystan around noon," Phil wrote. That was a distance of fifteen miles.

The crew moved into a cabin at Towdystan after stampede, a real luxury especially as it "rained like hell over the next two days." The tent had leaked and soaked Phil's bed. They had mechanical problems on the first of those days, and got the Cat stuck the next day. They put in a full day June 30, but Phil backed over Fred Engebretson's snubbing post. "I guess I'll have to fix it some day," he noted. July 1 they moved camp and back into the tent. Work was delayed when they went down the road to rescue Fred Engebretson who got stuck in the same place as the Cat. July 2 "was not a bad day," according to Phil. "There were lots of visitors which made quite a few stops. We also slipped the track today but we got it back on. Also a new cable, the other one came unravelled and wouldn't let the blade down." Graham did not get paid for breakdown time, neither did the crew. Phil recorded his work time, rarely less than twelve hours a day, sometimes fifteen. On the shorter days, he got the job of moving camp.

Betty Graham's husband, Fred Linder, arrived July 5 to take a shift on the Cat. Fred got the motor running backwards once but except for that, all went well. Fred Engebretson had a party that night and everybody went. On July 11, the district engineer came along and approved of the crew's work. It rained again. On July 12, the crew quit early to go to a dance at Anahim, but the affair was a flop. Everything went haywire—the record player went on the blink and all the lights went out. Howard Paley left the crew, he had to find his horses and get "squared around" for haying. July 15, it rained again, a good downpour. The crew was working in thick scrub jackpine. They just ran the Cat over top of them to knock them down, then pushed them sideways. Linder slipped a track the next day, which happened to be the first hot day. A rock jammed in between the sprocket and the tracks and threw the track off. One of the steering clutches was starting to go too. The sun finally shone in July, when the crew was headed home.

*Fred Engebretson. One of Chilcotin's notable characters, Fred was called the mayor of Towdystan. For years he was the only person living there. He was a renowned "haywire man"—he could fix anything mechanical. If he didn't have a part, or a tool, he'd make one. Except for fifteen years going to school and commercial fishing in Bella Coola, Fred spent his life at Towdystan, but even so, he probably knew as much or more about world and local affairs as anyone. He read a lot, listened to the radio, and had a chat with almost everyone coming or going to town. He was one of the first west Chilcotin ranchers to mechanize. He had to—he was allergic to horses. (Sage Birchwater photo)*

*Washout. Ted Gibbs stuck at Riske Creek, 1946.*

*Family travel. The Jim Ross family load up the car at Chezacut in the 1950s.*

*Alexis Creek, 1950s. The C1 Ranch House, built originally by Alex Graham, is in the foreground.*

They made Towdystan the first night, Caribou Flats the next. Phil saw a bear at Caribou Flats. He saddled up Paley's saddle horse and chased it, "hollering and whooping" but it wouldn't take a tree, and foiled him by crossing a swamp.

It wasn't all smooth going on the way home. A stick came up through the guard rails and bent the fan blade over into the radiator. Fred had to go to Colwell's to wire for a part. Cookie Hodgson arrived in the middle of the night with Brink's six-ton crawler tractor with blade and winch on his truck. The crew reached Clearwater that night, camped in a gravel pit near Dowling's the next night, and arrived home at Tatla July 26.

Lester Dorsey and his peers were right, and their worst fears came true. More and more people found Anahim as soon as the road got better. The late 1940s and early 1950s saw ever-growing numbers of hunters and fishermen invading the last frontier. When the Department of Indian Affairs built a school on the reserve that abutted downtown Anahim Lake, Southern Carrier people moved down from their remote meadows so their children could attend, adding to the village population. Mickey Dorsey, Lester's wife, was the first teacher. (She had no one to look after her youngest, one-year-old Frank, so she took him to school.) The road traffic increased as various government officials—game wardens, Public Works and Indian Affairs people—visited Anahim to check up on things. They needed somewhere to eat and sleep, it was too far to go back to Mackill's, and they put Ike Sing in the motel business.

The McInroys had cabins, but the government bureaucrats weren't about to cook or do for themselves. And they often arrived at odd hours. Ike didn't mind that, and feeding them wasn't a problem, but until he built some cabins, he had to bed his overnight guests on empty mail sacks in the post office. Some highly placed government officials slept on the mail sacks.

Ike had been getting by. He built a new store to replace the one that burned. He booked hunters and occasionally guided for Lester Dorsey. He worked on the road for Holtry. He took over the post office when Alyce Engebretson gave up on it. Alyce didn't mind the post office, but she couldn't handle the traffic that went with it. Anahim people lived far from each other and far from "town." When they came to get their mail every two weeks, they made a holiday of it. Most stayed for meals, some stayed for days. Harold, who was the telephone lineman, didn't have a lot of time for hunting and he was hard put to keep meat on the table. Alyce was busy with the ranch, and she didn't care for running a boarding house, so she was glad to get rid of the post office. The guests weren't a problem for Ike. Because he was "in business" it wasn't considered unneighbourly for him to charge them.

When Dowling left Anahim, there was no one to put on the stampedes. People missed them. Ike didn't know a thing about stampedes, but when one of his customers suggested he "make a stampede" he thought he'd try. Harold Engebretson and Holtry did know a lot about stampedes, and they helped. The event was a great success. The trio put on the next few stampedes and then the Stockman's Association took over. For more years than he can remember, Harold played the fiddle at Anahim Stampede dances. His sister, Thelma McInroy, who played the accordion, and her husband Earl, who played the banjo, made up the rest of the band. For some years Howard Harris from Quesnel joined them; later Alyce's brother, Jimmy Holte, played with them. Jimmy taught himself to play the guitar by listening to music on the radio, and he was very, very good. Lorena (Engebretson) Draney played the piano whenever there was one. Earl, who played melody rather than cording on the banjo, claimed the music came "out of a bottle" but whatever, the group would work all day helping with the stampede, then play all night at the dances, four nights in a row.

The 1950s saw seasonal fishing camps popping up on different lakes, including Anahim, Nimpo and Puntzi. When Alf Lagler left Anahim he built a store and cabins at Chilanko Forks, but most of the camp proprietors were newcomers, and most came for the season, then wintered in warmer climes.

CHAPTER 17

# The Highwaymen

*The road way is rough. Little graveling is in evidence. The few repairable chuck-holes are taking on a permanent appearance. What is disconcerting is the apparent lack of activity and appearance of maintenance crews.*
—Letter from Chilcotin residents, 1956

*The Alexis Creek mileage is 725 miles. We have two rotor graders and their maximum grading time is 170 days. That allows for two passes over the roads per year, providing there are no break downs. Maintenance, graveling and grading are cancelled out by the heavy lumber traffic.*
—District official's reply

In 1952, Public Works decided to have one foreman and one permanent full-time crew based at Alexis Creek to maintain all the roads in Chilcotin. Peter Yells was hired as the foreman.

Peter's father, Fred Yells, and a buddy named Oliver Handy had taken up land near Alexis Creek before World War One. Yells returned to England to join the army, and stayed there after the war. Peter came out to visit Handy, who had moved to Clinton, when he was seventeen, and he moved on to Chilcotin, working here and there, then joined the Canadian army at the beginning of World War Two. Peter must have heard the Drummer, because he returned to Alexis Creek after the war. He was working for Gordon Jakel when the road job came up.

Although politics played a big part when it came to road work in Bella Coola, it didn't in Chilcotin. Politicians tended to leave Chilcotin alone, probably because there were so few voters. Some weeks after Peter was hired, his wife Valerie was startled to see a notice advertising the job in the post office. She had some bad moments until she found out that's the way things were done. Local foremen hired whoever they wanted for crew, and the district engineer hired whoever he wanted as foreman, and it was posted after—probably to fulfill the legal obligations. The Alexis Creek job was offered to at least one other person, Roy Haines, but he didn't want it. Peter may have been asked because he was a veteran.

"I don't know a hell of a lot about roads," he said to Val after he accepted the post. She thought that shouldn't stop him. In the year they'd been married, Peter never showed any inclination to back down from anything. Val was new to Chilcotin, she'd come to work for Gan Gan Lee.

"I thought Roy Haines was foreman," she said.

"He was foreman for Bliss's territory, but now they're going to have one crew do everything from Sheep Creek to Anahim Lake and the side roads too. Roy says there'll be too much book work. He has that new grader and he'd rather stay with it."

"What exactly will you do?" Valerie asked.

"I'll be the boss." Peter whooped with laughter. Dark-haired, on the tall side and wiry, Peter wore a small, military moustache he'd grown in the army so he'd look older. His rowdy laugh was his trademark. He always laughed loudest when no one else saw the joke. A few others laughed when they heard Peter had the job—he was considered a bit of a newcomer—but Haines wasn't one of them.

"You'd think they'd get somebody with experience," he fumed to his wife Dorothy. "What does that young English ass know about roads?"

"Nothing," she agreed, "but I guess he's willing to do the book work."

"And take the BS," Roy growled.

Peter's first task was to tour the territory with Ray Cunliffe, the district engineer. Northern Construction was rebuilding the road between Williams Lake and Puntzi for the Canadian and American governments at the time, and as Public Works had nothing to do with it, Peter was free to concentrate on the west end of the road. Beyond Puntzi, it was two ruts heading west. Grass grew

between the ruts in places and there wasn't always room for
vehicles to pass.

"Why don't we just put a gate across it and a sign saying enter
at your own risk?" Peter guffawed. Cunliffe didn't think he was
funny. Given PWD's chronic lack of money and equipment,
Cunliffe agreed the only thing to do with the road was "gravel
the hell out of it." Peter scouted for gravel sites. His crew built
loading chutes, and attacked the road west of Kleena Kleene with
a Michigan loader and three dumptrucks. They didn't try to
smooth the ruts or pot holes, they just kept dumping gravel on
top of it all until freeze-up shut them down.

That summer, PWD built small cabins at strategic spots between settlements to house road crews and to provide emergency
shelter for travellers. The cabins were near water and wood supplies and each had a bed, table, chairs, stove, water bucket and
axe. There was one just past Redstone, others at Chilanko Forks,
Burnt Corral (east of Tatla Lake), Caribou Flats and Fish Trap.
People who used the cabins were honour-bound to leave cut
kindling and wood for the next guest.

Peter was expected to keep the entire Chilcotin Road open all
winter. Roy ploughed the lower end with the grader, Peter the
west end with the small bulldozer. It snowed early, and Peter left
on his first trip on December 26. He was to stay at the emergency
cabins. Hodgsons left fuel at each one, and Val cooked up a batch
of grub. There was absolutely nothing on the road. No traffic, no
animals, hardly any tracks, just miles and miles of lonesome snow.
The bulldozer's top speed was four miles an hour. Peter watched
the clouds scurrying by and counted telephone poles as he poked
along. He reached Tatla Lake December 31 and the Cat broke
down. He had dinner with Grahams and Bill Graham worked all
New Year's Day on the machine. Tom Chignell met Peter the next
day, he was going to swamp for him. Chignell drove a Model A
one-ton truck with a house on the back. It had a bed, a cut-down
wood stove and a radio, and he lived in it when he was on the
road working on the telephone line. The rig was the first camper
truck in Chilcotin.

The telephone line was down at Anahim and Chignell was
anxious to fix it. He and Peter got to Colwell's at Kleena Kleene

January 4. Gwen had turkey dinner with all the trimmings waiting for them. Also waiting were Alf Lagler and a passenger. Lagler's truck had slid over the bank at Brink's Lake five days earlier. The trio were staying with Adrian Paul, who took over the repeater station after the war. Paul was known as the Bird Man; he identified and recorded over two hundred species of birds around Kleena Kleene. Lagler's passenger had St. Vitus dance; his foot kept pounding the floor and it was driving Paul crazy.

The four travellers set off bright and early January 5. The sky was glassy blue, the snow glittering, the air knife-edged. Peter pulled Lagler's truck back on the road but it was chilled and wouldn't start. The chain kept breaking so Chignell suggested pushing it with the V plough. He dug around in his own truck and produced a can of ether which he gave to Lagler. "Put a capful or two of this on the breather," he said. Lagler unscrewed the lid and before the horrified Chignell could stop him, he emptied the can in the breather. Chignell leaped in his truck and took off down the road. Lagler got in his truck, planted one foot on the clutch and the other on the gas pedal, and signaled to Peter, who started pushing. When Lagler lifted his foot off the clutch, the motor caught with an ear-shattering blast and billows of smoke. It startled Lagler who slammed both feet on both pedals. The engine shrieked. Peter hadn't seen the ether episode; he thought the truck was blowing up. But before he could get out of the way, Lagler took his foot off the gas pedal. When everything calmed down, Chignell came back.

"Vat's de goot schtuff in de can?" Lagler wanted to know. Chignell wouldn't tell him.

Peter pressed on. Every inch of open ground between Brink's and Anahim Lake was enamelled with snow. The little Cat backed and bunted its way through the wind drifts. There was nowhere to stay between Fish Trap and Clesspocket, so Peter kept going, stopping only to catnap. It took him thirty-six hours to go twenty-five miles.

"You crazy young bugger," Andy Christensen said when Peter stumbled into the yard at Clesspocket at 9:00 p.m., bleary-eyed and wonky. Andy gave him a stiff shot of Scotch and put him to

bed, where he slept the clock around. The return trip was un-
eventful.

The next summer Pete's crew was back gravelling west
Chilcotin. Caribou Flats was particularly mucky that year and
everything bogged down in it. When a fellow came by on his way
to Bella Coola with a small International gas-operated Cat, Peter
hired him to pull vehicles through the worst spot. The cat would
waddle through the ooze, winch the car up, waddle back, letting
the cable go slack, then winch the car up and pull it.

BC elected the Social Credit government in 1952. Cariboo
people were staunch supporters of the Socreds, and Chilcotin
residents staunchest of all, but it was easier to push string than it
was to get money for any major road construction. Peter made
many little changes—"Minor Betterments" in government road
talk. He widened a corner here, made a detour there. Often a
settler's fence or field was in the way of a minor betterment, so
Peter bartered, begged and bullied for hundred-yard right-of-
ways, trading gravel or fencing or whatever he could. Chilcotin
people bellyached about the road but they really didn't care for
change. Whenever the district engineer came up with money for
anything major, like the Chilanko Diversion, someone would try
to stop it.

The stretch of road near Chilanko that Bob Pyper and Arthur
Knoll fought about was always a bone of contention. The grade,
known as Bear Head Hill (someone hung a bear head there once),
was standard Chilcotin—one truck width, steep and sidling. When
it was slippery, vehicles slid off into a deep gulch. If they were
lucky they could chop themselves out, but often they had to be
rescued. It was a ten-mile walk back to Chilanko Forks. Bear Head
was high on Peter's hit list, especially after logging trucks got on
it, and in 1959 he got some money to do something. He asked
Bill Graham to look at it. Bill went in by saddle horse and lined
out a twelve-mile route on the south, or Pyper Lake, side of the
hill. Peter approved. So did the district engineer, Dave McVicar,
and Peter hired Graham and Harold Stuart to build the new road.

People were upset. Evelyn Wilson, who had the store and
cabins on the old road, had reason to be worried. The diversion

left her business out in the cold. Tom Chignell was peeved because his telephone line followed the old road which was unlikely to be maintained once the new one was operational. Others who blistered the mails with complaints thought the diversion was a devious plot by Graham to get work for his bulldozer. Local wisdom said the old route was the best and just needed upgrading. Some wanted both roads maintained. Herb Coupe replaced McVicar in 1960, and he put an end to the letters, if not the tongues, by suggesting the road go through Chezacut, bypassing Chilanko completely.

People used the road of their choice and argued about which was fastest, until Stan Dowling and Anahim Lake rancher Bob Smith put it to a test. The two men were travelling west one winter evening, in separate vehicles, and at one of their stops they decided to settle it once and for all. "You take the new road, I'll take the old road, we'll both travel thirty miles an hour, and we'll see who gets to the top first," Dowling said.

Smith arrived at the junction first. He waited. And waited. Fearing the worst, he headed down the old road to look for Dowling. It was a cold night. Dowling had put a piece of cardboard in front of his radiator to keep the motor running warm, but the engine heated up on the unploughed road going up Bear Head, so he stopped to take the cardboard out. That only took a minute or two, and he was going along just fine when he met Smith, head-on, on the top of a rise. They hit hard enough to put Dowling's rig out of commission.

He always insisted the mishap was Smith's fault. "He knew I was coming. I didn't know he was."

Dowling got his truck fixed in Williams Lake, and was halfway up Sheep Creek on the way home when the fan went through the radiator. He put tobacco in it to plug the hole and drove carefully back to town. It turned out the repairmen hadn't noticed the housing cracked in the accident. Stan swore by tobacco for impaired radiators.

"It works and you always have some with you," he explained.

The Smith/Dowling incident settled the controversy. The twelve-mile diversion is now a permanent part of the planned

*Ike Sing. Thomas Squinas, who met Ike Sing on a hunting trip, talked him into starting a store at Anahim Lake, then helped him build it. The first store was built with mud-chinked logs and packing boxes. Ike, pictured here in the 1950s at his third store, always wore a cowboy hat and usually had a cigarette going. Ike had some lean times but he was in the right place at the right time when the Bella Coola road went through.*

*Tom Chignell and Gwen Colwell. These two had a working relationship that spanned many years. Tom established Halfway Ranch in the late 1920s, but he was best known as the west Chilcotin lineman. Gwen went to Tatla Lake to teach school, and following (or perhaps setting) a Chilcotin tradition, she married and stayed in the country. Her husband Sam was one of the first settlers in the Kleena Kleene area. For many years Gwen was the west Chilcotin switchboard operator. Whenever something went wrong with the telephone line, Gwen was usually the first to know, and when the line needed fixing, Tom was the one to do it. Here they are feeding some swans which somehow lost their way and ended up at the Colwell place at Kleena Kleene.*

*Chilcotin's first camper. Tom Chignell spent so much time on the road tending to the telephone lines, which frequently fell victim to heavy snow or falling trees, that he made himself a camper on the back of his truck. It was an all-weather affair, boasting a wood stove for cooking and heating.*

*Elgin Heath. When he first came to Alexis Creek, Elgin tried his hand at cowboying. The work was fine, but the pay wasn't, and when he was offered a job with the road crew he took it. He did every job there was to do on the road over the next twenty-plus years. No one will argue with his claim to have picked more rocks off the road than anyone else.*

*Stan Dowling. It didn't take many packtrips to and from Bella Coola to convince Stan Dowling that there had to be a better way to get supplies into Anahim Lake. Unlike most West Chilcotin pioneers, Stan didn't mind a bit of progress and he decided to truck in supplies. He saved enough money to acquire a used Ford truck in Vancouver, filled it with goods and brought it back to Anahim to open his first store. The dealer was glad to give Stan a reduced price on the truck in return for advertising how a Ford was the only vehicle able to make it over the rough track to Anahim.*

Chilcotin Road, and that seems to vindicate Graham's survey job.

Peter had his share of odd experiences. He was the only person to encounter an aircraft on the Chilcotin Road. He was going home one night, driving into the sunset, when he rounded a corner near Hanceville and there in the dust was a small plane coming straight at him, about two feet from the ground. Peter swerved and the plane hopped, but it was a near thing. The pilot, Bill Studdert, who owned Chilco Ranch at the time, was told to stop using the public road as a landing strip.

In the dozen years Peter was road foreman at Alexis Creek, a half dozen district engineers tackled the Chilcotin Road and they all sang the same song—the road needed upgrading, and the upgrading needed to be planned, not patchwork. Victoria replied with the same old song—the population was sparse, the traffic was light, the cost wasn't warranted. Chilcotin people didn't see it that way. They thought they deserved better, and as Peter was the closest target, he couldn't go anywhere without getting hell from somebody. He gave as good as he got, but he couldn't go to a dance or a wedding without someone getting on his case, and he grew tired of it. He finally had some cards printed to give out when he thought the complainer was out of line. The cards said simply "Bull Shit."

There is an old saying that "the weather's the poor's" because poor people can't move away from it. On the road crew, the weather belonged to the swamper because he was outside in it almost all the time. Swampers did everything. They drove the trucks that carried the fuel and pulled the caboose for the grader-men. They serviced machines, monkey wrenched, and changed grader blades. In the spring they gathered miles of snowfence and stored it away, cleaned culverts, and collected the tons of garbage that bred under the snow all winter. They usually did the cooking and they always picked rocks. From spring to freeze-up, swampers hoofed behind the grader with a long-handled fork (the tines were curved like a basket) and chucked rocks off the road with it. Thousands and thousands and thousands of rocks.

Elgin Heath did every job there was to do during his twenty-one years with the Alexis Creek road crew, but what he remembers most, not necessarily with joy, is the rock picking. It is a job you like less the more you do. It is safe to say Elgin picked more rocks and walked more miles on the Chilcotin Road than anyone else ever did or ever will.

When Elgin was in his early twenties, he lost a leg in a mill accident and he wasn't impressed with the artificial replacement. He thought his life was over until he visited his sister Margaret and her husband Doug Saunders who were working on the Telford Ranch at Alexis Creek. The Drummer's beat was just right for Elgin, and he was soon getting around very well. He worked for a time for Bill Bliss, and a few years later when Peter Yells offered him a job swamping for Roy Haines, he took it. He and Roy were together for five years.

Roy and Elgin were the odd couple. They had nothing in common. Roy, the son of a Riske Creek pioneer, was married with eight children; Elgin, a "newcomer," was a bachelor. They didn't like the same food, they didn't like the same anything. They shared a tiny trailer five days one week, six days the next week. It took them two years to work out mutually acceptable living arrangements. Roy and Peter didn't get along and Elgin was frequently caught in the middle of their rows, but, as Elgin kept telling himself, road work was the only decent paying job in the country so it was take it or leave it.

Alexis Creek had two graders at the time. Roy's territory was from Sheep Creek to Redstone, the area with the most traffic (up to 220 vehicles a day by 1960). Vehicles, especially big trucks, kicked up stones and debris as well as gigantic clouds of dust. Elgin learned to stand off the road when he saw a truck coming, and that gave rise to complaints he wasn't doing his job. The dust was as thick as flour and on hot days it stuck to sweaty skin like a horse blanket. Once Elgin literally worked in dust up to his knees—the grader got hung up on a rock that time, the dust so thick Roy didn't see it. Summer showers settled the dust but steady rain made it greasy, and a downpour made the road such a slop the grader had to quit. There was nothing to do but sit in the trailer waiting for time to go by.

Bugs were worse than dust. Wasps nested in culverts and attacked everything that moved, even the grader blade. Yellowjackets swarmed in angry clouds, and there were always mosquitoes.

There were no toilet facilities in the trailer or along the road, and every time anyone went into the bush, it stirred up the mosquitoes. All road workers were vulnerable, but the swamper usually got the worst of it—he had nowhere to hide.

There are several legends about the Chilcotin mosquito, and in one of them, Mosquito is a good guy. That can't be right. A Tsilhqot'in legend says Mosquito was a huge monster that lived on blood. He sneaked around at night grabbing people when they were sleeping and carried them away. The Tsilhqot'ins got tired of this and they devised a plan to stop him. They built a big fire, and lay around it, pretending to be asleep. When Mosquito came they all jumped up and pushed him into the fire. He burned nicely, turning into a big, black smoke, but then to their horror, the smoke started turning back into Mosquito. Somebody yelled "blow" and everybody started huffing and puffing really hard and they blew Mosquito to bits. Zillions of angry, bloodthirsty little bits that are still trying to get even.

When Roy retired, Elgin worked with others, including Delmer Jasper for three years. When Elgin retired, one of his gifts from the highways crew was a rock-picking fork.

CHAPTER 18

# Kings of the Road

*The big problem was the little gas pot trucks we drove had no power.*
*If you needed more power with a wagon and team, you hooked on*
*another team. You couldn't do that with a truck.*
—Trucker Don Widdis

The men who hauled freight and cattle between Anahim and Williams Lake were the kings of the road. They knew
all about the Drummer and Chilcotin time, even though many of
them lived east of the Fraser. They knew they would get where
they were going and back again, but they never knew when. They
knew adventure waited around each bend in the road, or over
the next rise, but they never knew what it would be. Like Chilcotin
people, they were pawns of the seasons, victims of the weather.
Distances didn't mean a thing, time was the only measure, and
everyone had plenty of time. If a trucker made a round trip
anywhere and back in breakup or winter without getting stuck or
breaking down once, he expected twice as much trouble the next
time. Given the potential for disasters, there were surprisingly
few. Chilcotin protected people who listened to the Drummer,
and the men who drove truck on the Chilcotin Road were good
listeners.

They were good drivers, too. Driving a truck wasn't just a
matter of sitting there until something happened. Like Chilcotin
people, truckers spent a lot of time in low gear, but they also were
constantly changing gears. Sometimes on the flats on a dry road
they could go a way without changing, but there weren't many
flats and they weren't often dry. They had to shift gears going
uphill, going downhill, in miry spots, on ice, in snow, wherever.
Faced with a challenge, drivers started in the highest gear they
could get away with, then geared down as they had to. All were
masters of the double clutch, a manoeuvre that required super

co-ordination and exquisite timing. If the truck stopped on a steep slippy hill or in a mud hole, it was all but impossible to get going again. Like everyone else on the road, truckers had to be innovative mechanics because garages were fewer and farther between than breakdowns.

Hodgson's was not only the first, it was the longest-lived truck line. The family stuck with it for almost fifty years. The boys started travelling with their dad when they were little fellows and they started driving as soon as they could get licensed. Jack and Wilfred (Its) were overseas during World War Two when Tommy had a series of strokes and Patrick (Cookie) took over when he was still a teenager. Roy Haines helped him hold the fort. Tommy died before the older boys got home.

The boys changed the company name to Hodgson Brothers, but they didn't change the nature of the service. No matter who was behind the wheel, the trucks delivered the mail, freight, passengers, packages, booze and news from one end of Chilcotin to the other, always without complaint, often without charge. Mail day was a big day in Chilcotin. There were always people waiting at the post office. Hodgsons kept to their posted schedule as well as they could, but nobody minded if they were late. They knew the truck would show up eventually. Alexis Creek postmaster Tommy Lee claimed that on cold winter nights you could hear the trucks coming a good hour before they arrived.

Dozens of men drove the Chilcotin Road for Hodgsons over the years. Roy Haines hired on for a few trips and stayed seven years. Vernon James (Mulligan) began driving before he had a licence, didn't get around to getting one when he was old enough, and drove for Hodgsons for several years before anyone noticed. The Drummer called Vernon, and when he left Hodgsons he moved to Chilcotin, bought property between Redstone and Chilanko, and married Lorraine Bliss, Bill and Irene's daughter. Vernon stayed with the road, though, driving logging truck and later driving grader for the highways ministry.

Other drivers who survived Chilcotin were Joe Gillis, who married Marge Hodgson; John Terrill (Teapot); Don Widdis (Squeak) who pioneered the Bella Coola run; and Barry McCue (Red) who bought the company in 1961. There were other truckers besides Hodgsons, including Gus and Gordon Jakel, Stan Dowling and Harold Stuart, and every ranch of any size had a truck to haul its own cattle or supplies. (Ollie Knoll's brother-in-law Ray Duggan was hauling a load of cattle one fall, and there wasn't much traffic, so he travelled mostly in the middle of the road. He didn't notice anyone behind him until he stopped for a rest, and an irate driver, who had been following in the dust for miles, leaped out of his car yelling "I know you god damned ranchers own the whole damn country but you don't own the whole damn road.")

There was no such thing as a typical trip on the Chilcotin Road; there was always a twist. Like the time Wilf Hodgson broke down on Tatla Lake. He was hauling to Anahim Lake, and had his first trouble at Redstone. An ice buildup by the bridge had backed water over the road where it had frozen into a thin shell. Wilf broke through it and the jolt snapped a spring. Jack brought another one from town but they didn't notice the drive shaft was bent. Wilf stayed that night at Mackill's, sharing a cabin with one of Chilcotin's better snorers, and got to Anahim the next day without further mishap. Tatla Lake was frozen solid and there wasn't much snow on it, so on the way home he took the shortcut down the lake. The truck whipped a bit on the ice as Wilf sifted along, and about ten miles down the lake, the drive shaft broke.

He hiked back to Graham's. It was a clear night, not too cold, but there was no moon. Wilf had a Labrador pup with him, and as they crunched along through the ankle-deep snow the puppy kept getting between Wilf's legs and tripping him. It made Wilf madder every step he took. He could see Graham's lights and hear the t'chug of the light plant, but he didn't seem to get any closer. When he finally got to the big house he was cold, tired and peeved at the puppy. The next morning when Bill Graham and Phil Robertson took him back to fix the truck, they saw the tracks where six good-sized wolves had followed him up the lake the night before. Wilf's neck prickled for weeks after whenever he thought about it.

More interested in mechanicking than in trucking, Jack Hodgson sold his share of the business to Joe Gillis, and he and his family moved to North Vancouver. Gillis was a giant of a man, standing six-foot-four, as strong as he was gentle. His truck lost its transmission on Sheep Creek once and several men were standing around trying to figure how to lift the new one in when Joe brushed them aside. "Get out of my way, I'll do it," he said, and he did.

Truckers carried whatever they could lift onto a truck and some things they couldn't. Gus Jakel once hauled a grain separating machine from Springhouse to Duke Martin's. It was too long to fit on his truck so he took it apart and borrowed a trailer from Charlie Moon. He knew he'd have trouble on Sheep Creek so he asked Don Mackay, who was ranching at Four Mile, to stand by with a team and chain to help on the sharp turns. As it turned out Mackay's team had to pull the truck around Cape Horn and by Moon's cabins and it took the better part of the day to get up the hill. Mackay only wanted two dollars but Gus thought it was worth more than that to walk the horses home. He gave him five.

Transporting chickens was a major pain for truckers. When Anna and Bob French moved to Alexis Creek from Riske Creek, Gordon Jakel moved their belongings which included a coop of chickens. Gordon stopped to talk to someone he met in Hance's Timber, and the chickens got loose. He spent the rest of the day running through the jackpine trees trying to retrieve them. Many Chilcotin residents bought baby chicks in the spring. Wilf Hodgson was hauling a crateful of them to Tatla one day when it turned very hot. Thinking the heat might get to them, he stopped and turned them loose on a sidehill. They took off in all directions; it took him hours to round them up.

Suppliers usually added a few extra chicks in case some expired along the way. When Joe Gillis delivered a hundred chicks to Anna French, she wanted to count them. Joe hunkered down and spread his hands to make a corral to hold the chicks. They counted 120 chicks—Anna thought she had a bargain—before they realized chicks were escaping between Joe's fingers.

Truckers had to be inventive to make a dollar. When George Christensen was building his lodge at Riske Creek, Dowling hauled a fair amount of freight for him. One trip he brought a boat up from Vancouver. It was bulky but light. Stan didn't know how to charge, so he brought 1000 bricks back too. He charged

Christensen $22 for the bricks and $100 for freight, then bought the bricks back for $22. He took them to Towdystan and sold them to Fred Engebretson for $45. Fred used them to build a chimney for his new house.

Truckers hauled passengers as well as freight—and "hauled" can be taken literally. Hodgsons and Dowling nearly always had passengers coming and going. As many passengers as possible crammed in the cab but extras had to ride in the back. If the truck was loaded, they rode on top of the canvas-covered freight. Tommy Hodgson called it the Hurricane Deck. In good weather it was pleasant up there; lolling on the canvas covering, rocking gently and watching the world go by as the truck ambled along the bumpy road. The sky riders could snooze, or visit if there were more than one. The Hurricane Deck was dismal in the rain, although Tommy Hodgson bought a bunch of buffalo robes from the BX when he left, and passengers could snug under them when it was cold. It wasn't easy scrambling up onto the freight, or getting back down, and the passengers squished whatever was right under the canvas, but no one complained. Riding in the cab wasn't all that wonderful, it could be hot and dusty and just as bumpy in there. Stan Dowling probably had the record for hauling the most people the longest distance—he took fourteen adults and a baby from Kamloops to Anahim Lake in his truck.

Until the 1960s, Native children in the Williams Lake District were sent to St. Joseph's Mission at 150 Mile House. Every September they were rounded up by Indian Affairs officials, loaded into the backs of the freight trucks like little sheep, and taken to the mission. They were returned in June. Wilf was taking the kids home one year, dropping them off at Toosey, Anaham, Alexis Creek, Redstone and Tatla Lake. When he got to the end of the line he had one little fellow left over. No one claimed him going back either. The boy didn't cry or kick up a fuss, but he stuck to Wilf like glue. Wilf took him to the Indian Agent when they got to town and it turned out the boy was supposed to be returned to the Quesnel area.

Gordon Jakel made one trip taking Dr. Hallowes to meet nurse Jane Bryant who was bringing a patient from Anahim Lake in a team and sleigh. They met at Clearwater. It was almost breakup and the road was awful. It took all day to get back to Alexis Creek. Not wanting to move the patient, they left him in the back of the pickup in Gordon's garage overnight, then took him on into town the next day. He recovered.

From the very first, Sheep Creek Hill was a challenge. It was too steep, and so narrow there was no room for error. Everyone hated it, truckers most of all. There was no such thing as guard rails and when it rained, or in the winter when it iced up it was treacherous going either way. If it rained, settlers on their way home preferred to sleep in their cars at the bottom rather than take a chance climbing the greasy track. If the road was slick, Hodgson's drivers didn't stop at Charlie Moon's place near the top of the hill in case they couldn't get traction to get going again, they chucked his mail sack out the window as they went by. If they thought it was going to be slippery, truckers on their way down would chop trees and drag them behind, butt end first, to act as a second brake. There was always a discarded forest at the bottom of the hill.

Once in a while someone fell off the hill. Stan Dowling was going down a bit too fast when he got on the soft shoulder and it gave way. The truck rolled completely over and landed on its feet in a gulch. Neither he nor his two passengers was damaged. The truck was slightly bruised but not immobilized. Luckily they landed on part of an old road, so Stan chopped his way back on to the new one. There were countless close calls. One New Year's Eve Wilf Hodgson was heading home in a Fargo truck with two passengers in the cab, one an older woman with a broken leg. When he stopped at Riske Creek to pick up the mail, there were twenty or so people from Toosey waiting for a ride to town. "Climb in the back," Wilf told them. There was no freight, just outgoing mail, so the passengers settled in with the mail sacks. Wilfred tied the tarp down snugly and away they went. It was early evening, and pitch dark. Sheep Creek Hill was glare ice and Wilf was descending cautiously, but he lost control on Cape Horn.

"The ass end of the truck swung out and I knew there was nothing underneath it because the truck was longer than the road was wide," he said later. "We spun right around and got going

*Hodgsons' farewell. Tommy Hodgson drove truck on the Chilcotin Road for over thirty years, and his sons for almost another twenty. When the brothers sold the business in 1962 they posed for this photograph. (L. to r.) Wilfred Hodgson, an office worker, Don Widdis, Vi Adolph, Lyman Pigeon, Willena and Cookie Hodgson, Marge (Hodgson) and Joe Gillis, Betty Jean (Hodgson) Johnson and her children Jean and Tommy.*

backwards. Then the back end hit the bank a whop and the front
end swung out. When it came around again I managed to ram it
into the bank."

When the truck stopped, the cab passengers were intact and
amazingly calm. Wilf got out, untied the canvas and stuck his head
in the back of the truck. He couldn't see a thing.

"How did you like that ride?" he yelled into the darkness.

"What ride?" somebody answered.

That stopped Wilf. It seemed they had been bouncing around
so much they hadn't noticed anything unusual.

Another time Wilf lost his brakes near the top of the hill, and
he geared down, but just before he got to Cape Horn the drive
shaft broke and he lost control. He was looking death right in the
eye when the front end of the broken drive shaft dropped and
dug itself into the ground, stopping the truck completely.

Tom Hodgson always carried a barn lantern with him. Wilf
wished he'd had one the night he stopped on the bridge to chain
up before tackling the hill. His flashlight kept wonking out and
made him mad, and when it quit altogether he chucked it over
the rail. On the way down it came to life and lay on the ice
winking up at him while he fumbled around in the dark.

Don Widdis started driving for Hodgsons in 1952. Don, alias
Squeak, was born in Woodpecker, a Cariboo community that no
longer exists. He drove truck from Vancouver to Quesnel on the
old Fraser Canyon Road and said nothing was ever as bad as that.
Most Hodgson drivers were hard drinkers and Widdis was the
cream of the crop in that regard. "I'll drive, you bartend," he told
his passengers. He claimed he drank anything that came out of a
bottle and Chilcotin home brew could be pretty rank. He smoked
four packs of cigarettes a day, slept erratically, ate even more
erratically when he was on the road, and by all rights he should
have ended up a basket case. Instead, he retired at sixty-seven
with a clean bill of health, still holding a Class A driver's licence.

Each week Squeak left Williams Lake around 5:00 a.m.—he'd
spend the day before going around town picking up the freight.
Each Hodgson's driver loaded his own truck. He'd probably have
way freight for Thompsons at the top of Sheep Creek, then for
Raffertys and Bert Roberts and Sawmill Creek. He'd drop over

the hill to Hance's post office, then cross the river to Chilco Ranch. Then it was back to Lee's and on to Alexis Creek. Squeak had a remarkable memory and if someone along the way wanted him to get something, they'd be sure to get it on the way back. All being well, he would get to Redstone around noon, Tatla Lake by suppertime, Kleena Kleene by late evening. There was very little traffic on the road, especially on the west end. In the winter of 1958, in six trips—one a week—the only vehicle tracks Widdis saw on the road between Puntzi and Kleena Kleene were his own. There was all kinds of game on the road then; one trip he counted forty-eight moose on Caribou Flats.

The truck went to Tatlayoko and Anahim Lake on alternate weeks but the schedule wasn't rigid, mainly because things didn't always go well. One breakup the high water in the upper country washed out the bridges on either side of Towdystan, trapping Widdis in between. He stayed with Fred Engebretson for twenty-eight days. It took something like that to stop him. Meals along the way could be rough. The drivers ate what, when and where they could. If they got stuck in winter it could be slim pickings because even if the truck was loaded with edibles, everything was frozen. Lunch could be a can of sardines thawed on the manifold. That could be supper too.

Under the circumstances it was understandable when goods and groceries didn't always arrive at their destination intact. Crackers squashed, bread squished. Sugar was a bummer, trucks didn't do much better than packhorses did with it. Wet flour got a hard crust but was fine inside, wet sugar either disappeared or turned to stone. Every west Chilcotin kitchen had a hammer and chisel for chipping flint-like sugar. Lumps were fine for tea or coffee but it took hours to whittle enough for a cake and often as not sparks flew. One spring Dowling hired Earl McInroy to meet Hodgson's at Kleena Kleene to pick up a load of sugar. "Take your time going but hurry back," he said.

The sky had been looking phlegmy all day and it decided to dump just as Earl got to Mackill's. Roy Haines had the Hodgson truck, and the sugar got rained on as he and Earl packed it from one truck to the other. Earl covered the sacks with a canvas. All was well until he got stuck on McClinchy Hill, which was steep, narrow and sidling, with a cutbank on one side and nothing for a long way down on the other. The road on top was made entirely of stones (a rocky bugger, according to the truckers), but the hill itself was powdery dirt that turned to slime at the first hint of rain. Every time Earl backed up and took another run at the hill he lost ground. He unloaded the sugar, piling the sacks on the bank and tucking the canvas around them. He jacked up the wheels and shovelled dirt from the bank under them for traction, but the dirt turned to mud faster than he could shovel. When the truck wouldn't budge in either direction, Earl walked five miles back to Pete McCormick's at Clearwater to borrow a team of horses. It was dark. His waterlogged clothes weighed a ton. It was too wet to roll a smoke and he'd lost his sense of humour by the time he got to McCormick's. He rousted Pete out of bed and declined his offer to help in the morning. He knew the sugar wouldn't last the night. Pete hitched up a team and Earl trudged back to the truck leading them.

In the meantime, Dowling phoned Towdystan, and Harold Engebretson and Tim Draney said they'd look for Earl. They hitched up a team and wagon and set out in the black, wet night. It was well after midnight when they found Earl, mud to the armpits, battling with the horses. Tim and Harold were top teamsters but they couldn't do a thing with McCormick's team even when they hitched all four horses to the truck. Tim was exasperated.

"They're just like Pete and his wife, they won't pull together," he muttered.

By daybreak they had the truck loose, downhill. They left it there. They returned McCormick's team and headed home in the wagon with the skinny bags of sugar. It was still raining.

Elton Elliott and Red McCue bought Hodgson Brothers in 1961. They ran it for eight years. Elliott Transport had twenty trucks in all, ten of them on the Bella Coola/Chilcotin Road.

CHAPTER 19

# The Bella Coola
# Connection

*AT THIS SITE ON SEPTEMBER 26TH, 1953*
*TWO BULLDOZERS*
*OPERATED BY ALF BRACEWELL AND GEORGE*
*DALSHAUG*
*TOUCHED BLADES*
*TO SYMBOLIZE THE OPENING OF A ROAD*
*THROUGH THE MOUNTAIN BARRIER OF THE COAST*
*RANGE*
*MARKED OUT BY ELIJAH GURR*
*TWO YEARS OF STRENUOUS LOCAL EFFORT*
*THUS ESTABLISHED A THIRD HIGHWAY ROUTE*
*ACROSS THIS PROVINCE TO THE PACIFIC OCEAN*
*THROUGH AN AREA ORIGINALLY EXPLORED*
*BY LIEUT. H. SPENCER PALMER, R.E., 1862*
—Plaque commemorating the opening of the Bella Coola
Road

$B$ella Coola residents refused to believe Destiny was
passing them by. Over the years the social clubs, political clubs
and special interest groups bombarded the government with
pleas and demands for a road "out." Their cries became louder
after Stan Dowling started his freight line from Anahim Lake to
Vancouver in 1934. The packtrains stopped going down the
mountain for supplies and the loss of Chilcotin trade put a
serious dint in the Bella Coola economy. About this time a new
voice joined the road chorus. It belonged to Clifford Kopas.

Kopas and his bride rode saddle horse from Alberta to Bella
Coola in 1933, and started a general store in the townsite. Kopas
was positive Bella Coola would take its rightful place in the

universe as soon as there was a road out of the valley. A prolific writer, Kopas wrote books, newspaper and magazine articles and letters promoting Bella Coola as the province's third outlet to the sea. In 1947 the Bella Coola Business Association was formed to promote tourism. He was elected president and doubled his efforts.

The Palmer Trail, the Rolston surveys, and the telephone line left the valley on the south side, but Bert and Peggy Matthews, who lived at Canoe Crossing, thought the Bunch Grass Trail on the valley's north side had possibilities. No one paid much attention to this idea until Norman Saugstad and Curtiss Urseth, top guns with Northcop Logging, took their wives on a pack trip over the route and came back agreeing with the Matthews. This triggered new interest in a road "out." The business association regrouped to form the Bella Coola Board of Trade and members focussed on getting a road up the Bunch Grass Trail. A government engineer who investigated the route pointed out that part of the trail leaving the valley was vertical, and he estimated the road would cost half a million dollars, but that didn't faze the Board of Trade. In 1930, residents had built six miles of valley road with every able-bodied man in the area donating a month's work and Public Works supplying the machinery. That project must have planted a seed because Board members talked themselves into building the whole road out of the valley the same way. Once started, there was no stopping them.

With $250 in the bank, they enlisted valley resident Elijah Gurr to find a route from Anahim Lake and to hire a bulldozer to build it. It took him a week to blow the bankroll, and over two years to build the road. For those two years he worked from dawn to dark, seldom taking a day off. He walked hundreds and hundreds of miles, wore out half a dozen pairs of boots and all of his helpers, seeking the best route for what the Board of Trade dubbed the Freedom Road. He also wore out himself. He was hospitalized twice with exhaustion.

Elijah, or Lige, was from Utah. In 1928 he married Isabel Meacham, whose parents were living in Bella Coola. The young couple visited them, liked the place, and returned in 1936 to stay. Lige was a logger, and he'd built his share of bush roads over the years, but he had been badly injured and was just getting his life together again. He was working as the school janitor when Board of Trade members approached him with their idea. Lige was a big, easy-going man who always saw a funny side in every situation. Once he started something he wouldn't quit. He was exactly the right man for the job. His first step was to take the road idea to Anahim Lake, where it was greeted with some disdain—and some laughter—by the Chilcotin Drummer's men. Lester Dorsey, Andy Holte and the other frontiersmen liked the country the way it was, without any roads. Storekeeper Ike Sing liked the idea. He knew Thomas Squinas was interested in a road, and he suggested to Lige that Thomas might help him find a route through the thirty-odd miles of muskeg and jackpine jungle between Anahim Lake and the Bunch Grass Trail.

"Thomas knows that country like no one else," Ike told Lige. "He traps, guides and hunts in there." Thomas had other priorities at the time but Lige was persuasive. Lester, who also knew the area well, took it upon himself to go too. They had no compass, but Thomas didn't need one, he always knew where he was, where he'd been and where he was going. He lined up Anahim Peak and Saddle Horse Meadow, then he and Lester rode along horseback, marking trail by notching trees with their axes.

In the meantime, Lige went to Tatla Lake to see Bill Graham. Bill was sympathetic. His dad dreamed of a Bella Coola connection and Bill did too, but now he did more than dream about it. He accepted Lige's proposal to start building the Anahim end of the road with his D6 bulldozer for $100 a day, to be paid when and if the Board of Trade ever got any money. He hired Alf Bracewell to run the machine. On September 12, 1952, the young rancher aimed the D6 at the blazed trees and began chewing his way through the bush to Bella Coola. Thomas had picked the high spots, avoiding the worst bog holes and ridges, and Lige refined it, slogging along a jump ahead of the bulldozer. The track went through brutal country, miles of pecker pole pine tree forest and spruce swamps so old and evil one expected to find primordial creatures slithering around in them. All of it swarmed with millions of famished mosquitoes.

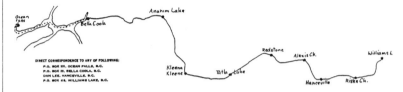

# CHILCOTIN-BELLA COOLA ROAD ASSOCIATION

**Sole Aim: A Good Highway-Car Ferry System from Williams Lake to Ocean Falls**

DIRECT CORRESPONDENCE TO ANY OF FOLLOWING:
P.O. BOX 131, OCEAN FALLS, B.C.
P.O. BOX 22, BELLA COOLA, B.C.
DAN LEE, HANCEVILLE, B.C.
P.O. BOX 116, WILLIAMS LAKE, B.C.

Bella Coola, B.C.
June 1st, 1963.

Mr. Buster Underwood,
District Engineer,
Department of Highways,
1890 Main Street,
North Vancouver, B.C.

Dear Buster,

A lot of people locally are getting their backs up because it looks as if we AREN'T going to get the black-top between Bella Coola and Hagensborg as suggested by yourself and promised by the Minister of Highways.

We understand that the delay is due to the break-down of a motor in the bull-dozer here.

I would like to have your assurance that road re-construction in our big and wealthy province is not being held up by the collapse of one only motor and that we will have the the black-top between Bella Coola and Hagensborg this year as promised.

Would you let us know as soon as possible so we can pass the good word on to an increasingly despairing populace.

Yours sincerely,

*Cliff Kopas*
Pres.

Road Association.

*The Bella Coola connection. In 1952 Alf Bracewell (on bulldozer blade, left) headed through the jackpine jungle west of Anahim Lake with Bill Graham's new D6 Caterpiller, building a road to Bella Coola. At the valley end, George Dalshaug (standing right) tackled The Hill with a geriatric International TD18. When the two met the following year it was cause for a great celebration–British Columbia's third outlet to the Pacific. Among those looking on was Elijah Gurr (seated on blade, right), foreman for the roadbuilding venture.*

*Official opening of the Bella Coola Road, 1955. The road was barely passable in 1955, but Bella Coola residents were using it so they staged the official opening. The plaque, engraved with the story of the two bulldozers meeting in 1953, was unveiled on this occasion but later stolen. L. to r.: Cliff Kopas, highways minister Phil Gaglardi, Mrs. Evan Jones, Evan Jones, unknown. (BCARS 39761)*

Bracewell was ten miles short of the valley rim in late October when winter chased him off the mountain. The roadbuilders still hadn't found a way down the 6000-foot mountainside. They abandoned the Bunch Grass Trail when they realized it would take fourteen switchbacks to get a decent grade. After more miles of walking, Lige suggested following Young Creek, a turbulent stream named for an upper valley settler. It begins in the Rainbow Mountains and tumbles its way down a steep, narrow valley east of the Bunch Grass Trail. Blondie Swanson, an engineer with Pacific Mills Northern Pulpwood in Ocean Falls, scouted the route, and agreed. With thirty-two miles of road under their belt and only ten to go, the Board applied to the government for a $10,000 grant. They thought the government might help because it had never rejected this new route, just the old one (several times). Their timing was perfect. Phil Gaglardi was the brand-new public works minister for the brand-new Social Credit government, and he was a roadbuilder. He was also a staunch free enterpriser who admired independence and initiative. The Bella Coola undertaking was right up his alley. Evan Jones, who earned his spurs in the Cariboo/Chilcotin country and was well acquainted with Bella Coola's ambitions, was deputy minister at the time. He was skeptical, but interested in the "new route." He knew it would take more money so he convinced the minister to set aside $50,000 for the project and to dole it out as needed.

With $10,000 in their bank account, the roadbuilders attacked the mountain from the valley side. To scale a rock face steep enough to give a housefly pause, the roadbuilders had a TD18 International bulldozer, well past its prime, operated by a nerveless young man named George Dalshaug, and a powder crew equipped with two air compressors and a couple of jackhammers.

Although Lige was the boss of the overall project, he was off in the bush most of the time. There was no foreman on the job, everyone just did whatever they were supposed to do. It all worked out.

As the construction crew bulldozed, blasted and chewed its way up the solid rock wall, dozens of valley residents hauled donated and scrounged supplies to the base camp at B. C. Wright's place near what was to be the bottom of the hill. There was quite a tent settlement there, at the foot of the mountain, as the families of the younger roadbuilders camped with them all summer. The main road in the valley was no joy at the time, especially when it rained. It deteriorated as the population along it dwindled—it was decent to Hagensborg, fair to Firvale, miserable beyond Stuie. The government's practice of responding to complaints instead of working to a plan resulted in some sections of road being rebuilt every few years while other sections weren't touched. "I remember driving through the same mess at Snootly Creek with a team and wagon as a kid, now I do it with a car. I guess that's progress," is how Alger Brinildsen summed up the road situation.

In December 1952, a snowfall of eleven inches was melted by a heavy rain within twenty-four hours, boggling the road good and properly. Then Snootly and Thorsen creeks broke out of their channels and cut the road. Somehow, someone always managed to get through with supplies for the mountain men. It didn't snow enough to shut down construction but the weather was shabby, and rain and fog made working on the exposed perch even more hazardous. Eagles watched in amazement, tall trees shook their heads in wonder, and clouds came down to supervise the workers as they pecked away at their prison wall.

The first $10,000 was gulped down in five weeks on the first mile and a half of road. Jones sent the second installment. Determined to get to the top no matter what, the crew took chances, especially the younger fellows who were as foolhardy as they were fearless. Everything was kind of haywire and it was a miracle no one was hurt. Dalshaug teased fate every day, skating over rock slides and crossing sheer stretches by anchoring the TD18 to large trees. The Cat slipped right off the mountain once. Dalshaug jumped free and the machine hung up on a rock outcropping about fifty feet down, stopping what would have been a thousand-foot drop to oblivion. They winched the machine back to safety.

There were all kinds of other frustrations. The blasting crew used old iron bits and changed them every two feet. One of the powdermen drilled for weeks on one section, working from dawn to dark, and when they finally blasted, the rock just lifted up and plopped back in place.

In the meantime, Lige looked for a way to connect the two

sections of road. They were only seven miles apart, but those seven miles were the roughest country they'd faced yet. Armed with an old aneroid barometer and an Abner level, both borrowed, Lige walked and scrambled back and forth, up and down, every day all day for the next few months. His final route reached a top altitude of 5200 feet at Heckman Pass on the Anahim side. It wriggled down Red Hill for 1800 feet to Young Creek (a 19 percent grade for almost four miles), clambered up to 4200 feet, then leaped off the mountain to the valley floor 3200 feet below. That summer the Graham cat was hired again and Bracewell had a go at the last few miles. Melvin Gurr was powderman and the team stayed in an old trailer that fell apart a bit more every time they moved camp. The stove smoked so badly Melvin would throw eggs in the frying pan, then run out to the fresh air while they cooked. Everything in the cabin was black with smoke.

They hauled the compressor in from the Anahim end by tractor but the road was such a muck after that they had to haul their supplies from Bella Coola. Jamis Jack was hired to pack the supplies up the Bunch Grass Trail. His horses were old, the trail was steep, and Jack was balky about hauling the dynamite, so when the two camps were within a mile of each other, Ole Nicola took over the supply route. He backpacked everything, including dynamite and fuel, up the hill from one camp to the other.

The two bulldozers were half a mile apart when the government grants ran out and Jones couldn't scrape up another nickel. Rather than stop, the roadbuilders agreed to go the jawbone route and finish the job on credit. In Bella Coola, Mike Christensen posted the road camp countdown, and on September 26, 1953, dozens of people drove up the valley to witness the meeting of the bulldozers. Four-wheel-drive vehicles ferried everyone up the mountain to the meeting place where Bracewell and Dalshaug were waiting to touch blades. Mike Stolar, the imported powderman, nailed his boots and hat to a tree, and left. The boots and hat were there for years. There were numerous toasts to mark the occasion, most straight from the bottle. A cavalcade of triumphant Bella Coolans drove to Ike Sing's at Anahim Lake where the celebration lasted several days.

When the dust settled, the Board of Trade was $8,740 short.

Every donor in the valley had been drained dry. The Board didn't know what to do, but Gaglardi was so impressed with their enterprise he bailed them out without even being asked. The forty-eight miles of road had cost almost $63,000 in cash, $59,000 of it government grants, $4000 from the community. That came to about $1300 a mile, not counting donations of time and materials.

It wasn't much of a road. What the Board of Trade had was a skinny shelf, a bit wider than a pickup, clinging to the side of the sheer cliff. It twisted its way for seven miles with three tortuous turns and a 20 percent grade—a climb of twenty feet every hundred feet. The rest of the road was navigable by jeep—if you had a very good jeep and if the road was dry.

In 1954, the government provided funds to "clear, grub and drain the road and do rock work on the big hill." Len Belliveau assumed duties as general foreman for the project but the public works foreman, Arthur Douglas, had to certify all the accounts. Belliveau wasn't used to red tape. His record keeping was a nightmare, and he wanted to use more PWD equipment than Douglas wanted to part with, so the situation was rife with conflict. Mobile camps were set up with Graham's D6 working on the Anahim end, a D8 from Bella Coola on the other end. This time, Bill Graham drove the D6 himself, and his wife Joy and young daughter Anne camped with him. His crew included Steve Dorsey, Lester's second son, and Ross Wilson, both from Chilcotin, and they spent the summer "clearing, grubbing and draining." The hills into Young Creek bristled with jackpines and the crew didn't slash a right-of-way, they just pushed trees over and piled dirt on top of them. Great sections of road were built entirely on trees. The ground was so hard they had difficulty scraping up enough dirt to bury them. Young Dorsey didn't know the word fear. He'd barge into the jackpines, knocking them down with gay abandon. Some of the trees fell over the bank with nothing supporting them but clear mountain air. Dorsey walked the Cat right out on them. Mrs. Graham kept expecting the trees, Steve and the Cat to slip into the chasm. Everything did go once, but the trees slid slowly and Steve got the Cat back on solid ground before they were gone.

Because the road was there, people used it. They were encouraged by the Board of Trade who wanted to prove it was needed, and discouraged by the road crews who spent a lot of time rescuing the travellers when they got stuck. There were some exciting incidents. A few Bella Coola residents entertained themselves by driving to Anahim Lake to spend weekends at Ike Sing's before Louis Svisdahl and his crew had finished building the bridge over Young Creek. One group, driving a van and a jeep, went up on a Friday, and when they came back Sunday the creek had risen. The Grahams were camped at Young Creek, and Mrs. Graham watched in horror as the van drove right into the creek and was swept away by the high water. It sank before it washed down the gorge, but the passengers had a scary wait while someone fetched the bulldozer to rescue them. Another jeep, trying to cross at night, was also swept away. It hung up on some snags. The occupants yelled loud enough to wake the camp but they got a good bouncing before they were rescued.

Andy Svisdahl was the first to try the new road with a passenger car. He told other people not to. Even four-wheel-drive rigs got stuck or slid off the road. One pickup shot right over the bank by Young Creek, a 60 percent grade, and landed in the road camp, narrowly missing the trees, tents and crew. Another fellow who went over the bank was asked by his rescuers if he needed a doctor. "I am a doctor," he replied.

While the Graham crew worked on Red Hill, Belliveau, an experienced powderman, was doing the rock work on the hill. One of his crew was a highballer who set off some healthy charges. If a stick of dynamite got stuck, he'd hammer it. He poked so much powder in one hole about three miles up the mountain it sheared off a section of the ledge, leaving a big gap in the road. Belliveau solved the problem by putting iron pins in the rock face and laying bundles of logs wrapped with cables across the pins. He covered it all with fill. Some years later the logs started to slide out, so Lige cribbed it with more logs. Eventually they had to blast farther into the mountain to made a proper ledge.

Melvin Gurr, the youngest member of the powder crew,

learned the true meaning of the word "cliffhanger" when he let himself be talked into going over the bluff to drill powder holes for the repair job. He was working from a tiny ledge, but he had a good toehold, and when his safety rope got in his way, he took it off. He was drilling away when the steel broke and kicked him off the ledge. He didn't fall far. He was right in line with the jack leg on the compressor and it pinned him against the rock face. There he dangled, 3000 feet above nothing, terrified someone would turn off the compressor and it would be goodbye Melvin. But the mountain gods were watching; someone got a rope down to him. Melvin took a couple of days off after the incident. "If it wasn't that it was something else, it was just part of the job," he said later.

Edwin Gaarden had the first of a number of poor experiences on the hill that year. Edwin's roots are deep in the valley: he is the grandson of F. P. Jacobsen, whose writing attracted the Norwegian colonists to Bella Coola. Edwin went to work for Arthur Douglas, the PWD foreman, in 1953, but he wasn't part of the Hill crew until the government became directly involved. He was sweeping rocks off the top switch with the PWD bulldozer, an old Allis Chalmers machine with a steering wheel, when the Hill tried to get him. He was head-on to the outside edge of the road when the pin fell out of the clutch and the Cat kept on going. For a hair-raising second Edwin thought he was going too but he managed to jump off. As he said later, the Cat had "lots of room to get going pretty fast" but all he could think of at the time was catching it on the second switch. It was particularly foggy that day. A crew working on the corner had stopped for a bite of lunch when they heard footsteps pounding down and young Edwin burst out of the mist. As he ran by he yelled "my Cat is over and I'm going to catch it." His colleagues were still pondering this when Edwin came back and said he couldn't find the machine. Everyone went looking then, but they had a time finding it in the fog. It had lodged against a stout tree between the switchbacks. When they winched the Cat back on the road, Edwin, who was sure he'd catch hell for losing it in the first place, insisted on riding it in case something went wrong. Something did. The cable broke and the cat went tumbling again, this time with the driver. It found the tree again and Edwin watched when they winched it up the second time.

The two crews worked from July to October, every day, with two compressors, a couple of hammers, a dozen or so people and one bulldozer at the Bella Coola end. The crews' families camped with them, and life in camp could be interesting. Whenever Belliveau was blasting, he would barricade both ends of the road and put up big signs to warn people. One day, just after he let off a large charge on the second switch, a Southern Carrier family with a team and wagon appeared out of the dust. The horses were pretty spooky. The man said, "rocks fell all over us." The crew's wives and children sat in trucks during the blasting and rocks fell all over them too.

Camping road crews could usually count on a bit of game or fish to brighten their menu of bacon and canned vegetables, but there was no game on the mountain. The animals probably left to avoid the mosquitoes which had a field day with the humans. No matter how many smudges the crews made nor how much insect dope they doused themselves with, they were bitten to bits. Ross Wilson, who was actually poisoned by the pests, spent several days in the Bella Coola hospital.

The road wasn't without its critics, and one of them was Belliveau. "He thinks it is wasteful of public money to attempt to keep the present road in repair," Kopas complained to Gaglardi in September. "He wants to reroute and straighten sections of it to bring the whole thing into condition to be built into a highway." The minister ignored the suggestion.

In December 1954, Lige Gurr was appointed Public Works foreman for Bella Coola. He replaced Douglas, who had held the job for almost twenty-five years, a remarkable feat in itself as Bella Coola has had a reputation for being hard on road foremen. In April 1955, the embryo road proved its worth when a strike stopped the steamship service for two months. Logging trucks were pressed into action to haul food and supplies from Williams Lake to the stranded valley.

The official opening of the Bella Coola Road took place on July 17, 1955. Gaglardi was there for the occasion, flying to

Nimpo Lake where he was picked up by Evan Jones and his wife, who drove up from Victoria. Jones was driving the official car, a big Buick. There was a heavy rain a few days before the opening and the Kleena Kleene river slurped over the road by Stan Dowling's. The water was deep but the road underneath was solid, so Peter Yells made a deal with Dowling to pull vehicles through the flood with his tractor for $5 a shot. Dowling put up a sign "Free Tows" and spent hours pulling cars across the flat.

Mrs. Jones wasn't expecting this. "Is my suitcase safe in the trunk or will the water get in and ruin my clothes?" she asked Dowling.

"I don't know," he replied. "I haven't looked in the trunks of the cars I've pulled through."

The westbound vehicles, an assortment of trucks, cars and jeeps, proceeded from Anahim Lake in a caravan. The rain had mushed the entire road and vehicles bogged down in mud holes, hung up on roots and rocks, and slid off the road into the bush. It took the combined manpower of the entire convoy to keep the show on the road. Every vehicle had to be assisted through one spot near Anahim. Low-slung cars took a terrible drubbing. The drivers and passengers were well supplied with liquid refreshment to fortify their spirits, and a party mood prevailed as they schlumped along.

The ribbon-cutting ceremony was held at the spot where the two bulldozers met in 1953. A plaque marked the site until somebody stole it.

Outsiders saw the road as a goat trail cleaving to the mountainside and crawling its way through the jackpines. (At the official opening, Gaglardi was heard to comment, "This opening is a bit premature.") To the Bella Coola residents, it truly was the "Freedom Road." Where it cost $250 per person for return fare to Vancouver on the steamship, which made the trip once a week, a carload of people with enough nerve could make the trip one way in three days for $30 worth of gas.

But the road did not bring the prosperity the Board of Trade expected. Many tourists who tried the road shipped themselves and their vehicles out on the *Northland Prince* rather than face a return match. Few returned for seconds. Chilcotin and Williams

Lake residents considered it a badge of courage to make the trip, but once was enough for many of them, and they didn't add much to the valley's economy. The main traffic going in was either Anahim residents or fishermen who wanted to try their luck in the river.

The Board of Trade didn't give up. Its stationery included an impressive letterhead proclaiming "The Third Outlet to the Sea." Members thought the province should connect their road to Prince George to fulfill that city's destiny. Kopas kept writing letters and articles and was a one-man Chamber of Commerce until he died in the late 1970s.

CHAPTER 20

# On the Road

*No trouble to Anahim Lake, 225 miles, the first 190 in excellent
condition. The rest is passable depending on the season and
amount of rain. The Bella Coola Road is a road in name only.
We suggest you get in touch with the Bella Coola Board of Trade,
the father of this road.*
—Williams Lake District office, 1954

Chilcotin pioneers were grateful for road work, first
because they could make a dollar doing it, and second because
it improved their access. With the establishment of permanent,
full-time road crews in the 1950s, the job potential was lost, and
as traffic increased, so did complaints. Residents complained
about the road, about the road crews and about each other. Some
people worked hard to keep the road crews honest. Foremen got
the worst of it. Grader operators came second. They were called
Feathertough, Neverscratch or Highblade. People thought the
crew should be on the main road all the time, forgetting there
were hundreds of miles of side roads. Members of the crew didn't
always get along with each other, either.

There was no snow to speak of in the winter of 1960, so Pete
Yells sent a crew to Tatla Lake to put punchin (poles laid cross-
ways) in a stretch of road, actually a lane, between Graham's
hayfields. Jim Mackay, Bill Telford, Bob French, Sam Martin and
Charlie Cammiade were there for three months. They used
power saws to cut and limb jackpine trees into about 10,000
fifteen-foot logs. Tex Hansen, a retired boxer who had a place at
Clearwater, skidded the logs out of the bush with horses. He was
a bit absent-minded and kept forgetting where he left his spare
team. Punchin laying is labour intensive. The crew had a loader
to put the logs on the dump trucks, but they had to lay them in

the lane by hand. The trucks hauled fill, and the grader buried everything.

The fellows lived in the cabin at Burnt Corral. They worked long hours because there was nothing else to do. In those days they travelled on their own time, so they worked ten days and went home for four. Since it took the better part of a day to get home to Alexis Creek from Tatla, they were only home for two full days at a time. Even the best of friends got on each other's nerves under the circumstances.

Mackay, an antique car buff, had with him some spoke wheels from an old Buick which he worked on in his spare time. They took up quite a bit of room in the cramped cabin, as did Cammiade's dog, a purebred golden Labrador which slept by his bunk and ate from the table. When someone shot a deer, the dog had a big feed of raw meat and got gassy. It really stunk up the cabin, and this bugged everyone, especially Martin. One night, when Cammiade went to Tatla Lake, Martin found some orange paint somewhere and painted a stripe down the dog's back. Everyone was in bed when Cammiade came home. The dog was outside, and the truck's headlights caught the paint job. Cammiade burst into the cabin yelling "you painted my dog," and went straight for Martin. They weren't big men but they made a big commotion crashing around and beating on each other. Mackay played possum. Telford and French tried to stop the fight, but the cabin was totally dark and they couldn't see who was what. They tried to light the coal oil lamp but the fighters kept tumbling into them. When the Buick wheels got loose, rolling around and whacking everyone, they gave up.

"Maybe they'll play themselves out before they kill each other," Telford said. They did. The cabin was a terrible mess in the morning. So were the fighters. Cammiade was mad at all of them; he thought someone should have protected the dog. It was the only time Telford and French, both of them well over six feet tall, couldn't stop a fight. "We were bested by the Buick wheels," Telford claimed.

Road crews did some construction work, but mostly they graded and gravelled between breakup and freeze-up. In winter

*Stuarts at Redstone. In the early days Andy Stuart minded the store at Redstone with his wife Hettie. When their daughter Christina finished school, she ran the store and post office while her brother Harold drove truck and ran the garage. Christina died in the 1970s and Harold's wife Marcella, pictured here with Harold, took over the business. In the late 1980s, when Harold died, Marcella closed the store and moved to Williams Lake. Photo by Don Wise.*

*Rock picking. Picking rocks off the road was never a fun job. For every rock chucked off the road, another three seemed to sprout. The road on top of McClinchy Hill was a "rocky bugger" even by Chilcotin standards. Bob Smith of Anahim Lake, best known for his bronc riding at Anahim stampedes, is shown here with Bill Graham's crew in the 1940s.*

*Washout, Alexis Creek, 1962. In springtime, when the snow melted and frost came out of the ground, the road flooded wherever creeks overflowed or culverts couldn't handle the runoff.*

*Anahim store, 1960s. Ike Sing sold his business in 1960 to Ed Escott, a retired tugboat skipper from Vancouver. Ed's daughter Marilyn and her husband, Don Baxter (above), later bought it. Like all Chilcotin stores, the Anahim establishment carried an eclectic stock, catering to the needs of the local people and also to hunters, fishermen and any tourist who happened by.*

they ploughed and sanded. Some winters they spent a lot of time fighting with the ice that backed up against the bridges, taking them out or backing water over the road. A good shot of dynamite usually fixed the ice. None of the crew were trained or licensed powdermen, but that didn't stop them. When the Kleena Kleene River froze behind Chignell's place, French and Telford got a box of dynamite from Alex Graham at Tatla Lake. He didn't want it back because it was old and weeping, so they used it all in one shot. The blast took out the ice and some windows in Chignell's house.

When ice threatened a new bridge at the Newton place, the crew rigged up a stick of dynamite with a rock so it would sink and blow under the ice. Their timing was off and it blew under the bridge, causing some damage and a lot of red faces. The first time Jim Mackay got hold of some dynamite he used the whole case to blow one tree. "Do you think it will blow it?" he asked. Everybody ran. The blast blew the tree right over the telephone line and it landed intact in the middle of the road. Another time they blew the telephone line out.

In 1954, road work was separated from Public Works to become the Department of Highways (DOH). Chilcotin continued to call it PWD, or Public Works, and the crew public jerks. Soon after, government employees formed an association, and the leaders called a one-day strike to show the government they meant business in asking for better pay and better working conditions. The idea of a strike was totally foreign in Chilcotin, where people were used to taking what they got until something better came along. The Alexis Creek crew obediently stayed home on strike day, knowing God would get them for it. Peter Yells, being management, sat in his little office all day feeling lonesome. Everyone took the better pay when it came, and the benefits the association negotiated. They liked the shorter hours and the pension plan, but they didn't understand the concept of organized labour. It didn't march with the Drummer.

The year 1962 started with the Big Snow. It began falling on a Saturday morning in late January and came down steadily for two days, great goose-feather flakes of it. By Sunday night the entire central Cariboo was under forty-two inches of snow. No one had seen anything like it. Everything was stopped cold. Road crews were on the job right from the beginning but they couldn't keep up. Phone lines were out so there was no way to communicate. Everyone just kept working until the roads were open.

When Red McCue left Williams Lake Tuesday morning with the Chilcotin mail and supplies, the road was open as far as Alexis Creek. French and Mackay were west somewhere with the grader, no one knew where. McCue went on, hoping for the best. He made the first ten miles in half an hour, the next ten miles in three and a half hours. He kept going. French and Mackay were at Brink's when the snow started. Mackay was swamping–driving the service truck which also towed a little house trailer. They stopped only to catnap and didn't have a hot meal until they got to Chignell's on Tuesday. They camped on Pyper Hill that night, and when they started up at 5:00 a.m. Wednesday morning, they saw McCue's lights at the bottom of the hill. He got to Anahim Lake at 8:00 that night after twenty-six hours of steady driving. French and Mackay made it home that night.

French moved to Anahim Lake that spring, and Yells went to Williams Lake the next year. Telford took over as foreman at Alexis Creek. The lower Chilcotin Road was wider and straighter than the west end, and there were no serious mud holes, but by the mid-1960s it was completely worn out. Little money was spent on any of it. French and Telford made do with Minor Betterments. When Telford wanted to straighten a piece of road or detour a mirey spot, he had to make deals with property owners for bits of hayfield. He did get $120,000 to rebuild the road from the top of Sheep Creek to Farwell Canyon. It was barely enough to cut off the bumps and a corner or two and widen the road a bit.

There were few ranches near the road at the Anahim end. French just had jackpines, swamps or boulders to contend with when he widened a corner or skirted a mudhole. He had different kinds of problems, like the time a culvert on a creek near Clearwater backed up, causing a minor flood. When the water went down, French found a moose hide tied inside the culvert, obviously put there by someone soaking it for buckskin. He hung it on a fence and it disappeared a few days later.

The Bella Coola crew did their best to keep the new road to Anahim Lake passable. In the summer of 1955 Lige Gurr got a new grader and bulldozer, and there was a big push to improve the road between Anahim and Saddle Horse Meadow. Some of the crew had their families with them. They camped at Green River, a favourite hangout for mosquitoes. Lige had five trucks hired but he couldn't find any decent gravel. The road was soup; it took a Dodge power wagon four and a half hours to drive the twenty-one miles to Anahim. It was thundery and showery all summer and travellers kept getting stuck, nearly always at night. When the crew rescued them, local people and hunters always said thanks with a drink, sometimes with a bottle.

They made a noble effort to keep the Freedom Road open in the winter of 1955 but their equipment wasn't up to it. The little bulldozer with its dirt blade and the old grader could only clear a bicycle trail. Altogether they spent six weeks on top. Edwin Gaarden said it seemed like six months. They worked with two-man crews and took turns sleeping in the truck. When they did see a bed it was in a tent that kept collapsing under the snow. At East Branch they ran into five feet of sugary snow that fell back on the road faster than they could push it off, so they gave up.

The road wasn't that great in the summer time, either. Even Lige was cautious. "It is very difficult for low cars but jeeps and trucks are OK. We do not encourage unnecessary traffic over the road yet, but it is passable in the summer for those who drive carefully," he warned in an official report in 1956. They tried to keep the road open that winter too. This time they had an airtight heater, but it melted the snow under the tent. They put boughs underneath and moved the tent every other day, but it dripped steadily and their beds were always wet. The temperature dropped to -45°F in January and chased them off the mountain. The road wasn't open again until March.

In 1957 Lige acquired an old trailer, a caboose. It was a big improvement over the tents. In the summer the crew took a truck with barrels of fuel up the mountain and towed the caboose. As well as gravelling and widening the road, they had to replace the hundreds of jackpine culverts put in by the original roadbuilders. It took years to do that.

In the winter of 1957 Lige opened the road so people could go out for Christmas. Holidayers were escorted by the bulldozer on specified days before Christmas and after New Year's. Mack and Dean Gurr were the escort service the first year. They were pulling the caboose and it was slow going. Some people left Bella Coola, thinking the road was open, and caught up with Gurrs at East Branch. They had to trundle along behind the TD18. Gurrs had food for themselves, but their four guests had nothing, and the food situation was tight by the time they reached Saddle Horse Meadow. To make better time, Gurrs hooked the vehicles behind the trailer and pulled the whole lot. The caboose kicked snowballs the size of sheep at the towed vehicles so they abandoned it, but they hadn't gone far when the car of one guest quit. The snowballs had rolled under it and squashed the gas tank. The needle showed half full. When the convoy reached Anahim, some people had been waiting at Ike Sing's for a week.

The Bella Coola connection was a bonanza for Ike. He'd built more cabins as the need for overnight accommodation grew. One cabin was the height of luxury: it had a chemical toilet that flushed on a quart of water a day. He also built a bath house which was popular with travellers who arrived at Anahim caked in Chilcotin. Ike had a helper, a Mr. Yee from Vancouver, a wonderful cook who could whip up a tasty meal in minutes from practically nothing. Ike replaced his first little gas-operated plant with a big diesel generator, and he kept both in the bath house building. The Drummer must have thought he was doing too well, because he gave him another poke in the eye. Ike was driving home with a load of freight one cold stormy night when he played out at Fish Trap and took shelter at the emergency cabin. He never knew exactly what happened, but Mr. Yee somehow burned the bath house and both light plants. Ike was close to the end of his tether that time, but he started over again.

Ike sold the business to Ed Escott and his family in 1960. Escott had a towboat business at the coast but he hunted and fished in the Anahim country and was hooked by the Drummer. His daughter Marilyn and her husband Don Baxter were to be Mr. and Mrs. Hospitality at Anahim Lake for the next thirty years.

Originally, the Bella Coola Road went through the edge of the

Indian reserve at Anahim Lake. It actually went through someone's homesite, so it had to be moved, and when it was, it bypassed downtown Anahim Lake. In the early 1960s, downtown Anahim Lake was the Escott store and cabins, Andy Christensen's store, the Frontier Cafe and McInroys' garage and cabins. There were some good rows as to where the road from the highway should go—in by Escott's, or in by McInroy's. One night the bridge on the McInroy entrance mysteriously burned. Everyone blamed everyone else and it was years before the culprit owned up. It was Lester Dorsey. "I thought I'd add a little fuel to the fire," he said.

Lige went to Williams Lake in 1959 as senior foreman, and his son Mack took over the Bella Coola crew. The road was kept open in winter on a hit-or-miss basis until Williams Lake established a maintenance depot at Anahim Lake in the summer of 1961. The depot itself operated on a hit-or-miss system at the start. The first foreman quit after three months. His replacement was fired after he left the bulldozer and the grader on the mountain where they got snowed in. Bob French took over in May 1962. The depot was a large garage, a bunkhouse (the Caribou Flats emergency cabin) and the foreman's house, a fort-like structure built of vertical logs. The depot sat in the jackpines half a mile from the village. Frank Dorsey, Lester's seventeen-year-old son, operated the aged D7 Cat. The equipment included a brand new dump truck, a newish grader, a geriatric tractor with a bucket and a pull grader. The French family included two school-aged sons who brought the off-reserve child population to ten, enough to open a public school. French, who was born at Kleena Kleene and operated the grader at the Anahim end of the road for six years, was well acquainted with the area and thought he knew what he was getting into.

Anahim Lake was in the Williams Lake District, Bella Coola belonged to North Vancouver, and no one was sure where the boundary was. French's crew went to Green River in summer, East Branch in winter, and while they had snow dumps of three feet at a time to contend with, the Bella Coola crew had the Hill. The installation of the radiophone system in 1962 helped everybody keep track of each other. The base sets were in the

*Telephone office. For years this building was the home of the Cliff Kinkead family at Alexis Creek. It also housed the telephone communications system for the Chilcotin Valley. Nellie Kinkead was in charge of the switchboard. Cliff was lineman for the Dominion Government Telegraph Service, later taken over by BC Tel. Vera Hance, a longtime operator for the valley's telephone system, is shown below in 1965. A dial system was installed later in the 1960s, eliminating the need for operators, but the building is still there, housing a gift shop and other services.*

*River road. The Bella Coola Road in 1968. The river often flooded this low stretch near Firvale so it was "rip rapped"–large rocks were put along the bank to keep the road from washing away. Photo by Art Long.*

foremen's homes at Alexis Creek, Anahim, and Bella Coola, and the foremen's wives were vital links in the system. The Anahim base set reached Alexis Creek and Williams Lake on a repeater, but the Bella Coola–Anahim connection was direct and exclusive.

Ironically, the radio system was responsible for the only crew deaths in Chilcotin. The radio technician, a young Dane named Ole Herring, was based in Williams Lake and made regular trips west to service the system. He was the first person known to drive from Williams Lake to Anahim in five hours (the previous record was six). The repeater was located on Puntzi Mountain, and when it went out in the winter of 1964, Herring and Ray Hubble from the Williams Lake crew snowshoed in to fix it. They lit a camp stove to warm their supper, not realizing the little building that housed the repeater was airtight, and they died from lack of oxygen.

Before the radios, the only way to communicate with a crew on the mountain was in person. The first winter Mack Gurr was foreman, he went to check on the mountain crew after he hadn't heard from them for a week. About six miles up the hill, the clutch went on his pickup. It was ten miles back to a phone, so Mack opted to walk the seven miles up to the road camp. The temperature was falling. Mack's heart fell too when he found the air inside the trailer eerily cold and the propane tank that fuelled the stove and heater missing. He guessed the crew had pushed on to Anahim Lake. There was no way to light a fire.

Gurr found some frozen sausage and chewed on that, then bundled himself up in all the sleeping bags. He'd worked up a sweat walking up the hill, and his thermal underwear didn't absorb it and was cold and clammy. He was almost too cold to care when the crew came in around 10:00 p.m. with a full propane tank. They'd been gone for two days. The temperature hit -60°F that night and they had to leave the machines running.

Even with the radios there were communication problems. The crew didn't go near the mountain unless they thought it was safe, but it fooled them once in a while. One night in 1963, grader operator Gordon Levelton was on the big hill, near Cougar Rock, alone, when an avalanche caught him. The grader was equipped with a wing for snow ploughing, it stuck out about three feet and the road was barely wide enough for it. When the wing was on the bank side, Gordon had to be careful the grader wheels didn't go over the outside edge. When the slide hit, it buried the grader and pushed the windows in. He sat in the machine, afraid to move, listening to his heart pound, for about half an hour. He tried to get Gurrs on the radio but he was in a dead spot. Finally, when nothing else happened, he dug himself out and scrambled over the slide. He was on the second switch and the trailer was at the bottom of the hill. It was spooky walking down in the dark. He could smell the fresh dirt disturbed by the slide and he could hear plopping noises behind him. He wasn't sure if it was snow or a cougar stalking him, but there was nothing he could do except keep on walking. He was soaking wet from crawling through the snow, and when he lit the propane stove in the trailer it steamed up so much he had to open the door. There was a phone in the trailer so he called Mack, then went to sleep. He didn't look at his watch, but when Mack woke Gordon's wife to tell her what happened, it was 5:00 a.m.

In the winter of 1965 there was a spell where it snowed, got cold, then warmed up. Mack and another Cat operator were at East Branch, and they thought they could get home before there was any trouble. About three miles out Mack felt the tell-tale wind on his face. He stopped the Cat, intending to climb under it, but the avalanche hit before he could, and covered the Cat. Mack dug himself out just as the second machine came along.

Every year there were dozens of little slides between the foot of the Hill and Young Creek. When it rained, the ploughed road on the Bella Coola side of Young Creek got so slick bulldozers couldn't walk on it. The icy conditions lasted ten hours or so, and every so often the crew got stuck on the wrong side.

Don Widdis took the first Hodgson's truck over the Bella Coola connection in June 1956. The next year he went once a week between breakup and December. Widdis had some unique experiences. Once he was going down the hill when a rock broke an axle and his brakes failed. He geared down but another rock broke the two-speed gearshift and the truck was freewheeling. His two passengers were pop-eyed by the time he got it under control by ramming it into the bank. Widdis's truck always had bashed fenders from bunting into the bank to stop or slow down. In winter he chained up all around, and it was nothing to go through a set of chains a trip. Once he was stuck for two days on a sharp turn on the mountain because his chains were so wrecked he couldn't even wire them together. Another time the road washed out and he was stuck in Bella Coola for several weeks. He took a job down there.

Someone once asked Widdis about his worst experience on the Hill. He thought for a minute then said nothing was as bad as meeting Alf Lagler on Sheep Creek. He started his own company, Widdis Trucking, and hauled into Bella Coola for a time, and then he headed south to warmer climates and better roads.

When Red McCue and Elton Elliott bought Hodgson Brothers in 1961, Red took the Bella Coola run. A good part of his weekly freight was liquor, cases of it, because the Norwegian colonists' constitution prohibited any liquor outlet in the valley. Every few years some group or another would sponsor a plebiscite to change the rules, but the dries always voted it down. On his trip down before the 1963 Fall Fair, McCue had three tons of booze on board. It eventually dawned on someone in Victoria that the colonists' constitution was hardly binding seventy years after the fact, and a liquor store was established without any vote.

In winter McCue would honk his horn when he went by the Anahim highways yard, and again when he went by Gurr's in Hagensborg. On the way back he reversed the procedure. If he didn't arrive at either end in a reasonable time, someone would go looking for him. He didn't get stuck often, but when he did it was a dandy.

In the winter of 1966, the Bella Coola bulldozer broke through the Young Creek bridge while Red was in the valley. He was driving a five-ton Ford, and when he came back he drove across the creek, which was full of ice. He couldn't get up the bank on the Anahim side. He knew Steve Dorsey and Butch McInroy were nearby, so he walked until he found them. He and Steve drove back to the truck and they were having a coffee while they waited for Butch. Red suddenly yelled and ran for the truck. The ice had built up against it as high as the cab and it was moving. They tied the truck to the trees with Red's logging chains, but there was nothing else they could do but watch as the ice built up under the truck and rolled it over. When it was on its side, the ice shot over the top and buried it. When Butch arrived, he pushed the ice away, pulled the truck back on its wheels and up the bank. The frame was bent, but once they chunked the ice out of the motor it started, and Red was on his way. If he hadn't been quick enough to chain it, the truck would have taken the bridge out and gone down the canyon.

Dorsey had his moments on the Hill. Raised at Anahim to the Drummer's beat, he spent most of his adult life in Bella Coola. Like other frequent travellers, he had a few tricks for the Hill. When ice built up on the bank side, he chopped a narrow track, eight to ten inches deep, to hold the inside wheels. A skim of ice could be roughened up by throwing gasoline on it and lighting it. He backed down the hill once. He had a standard transmission and good V-grip chains. He figured the chains would chew him to a stop, and he could butt the bank if the front end slipped. It worked, but he had to chop his way around the second switch.

The mountain road has no secrets from Edwin Gaarden, nor do any of the roads in the valley. Edwin spent thirty-seven years with the Bella Coola road crew. He had more battles with slides, deep snow, floods, mud and mosquitoes than anyone else had or is likely to. He was seventeen when he started driving dump truck for a gravelling crew. They had nothing to spread the gravel with; if the truck driver didn't do his job properly, it was rake and shovel time. Edwin learned very quickly. He had his eye on the little bulldozer which was foreman Arthur Douglas's pride and joy. In those days it was every man for himself when it came to learning to run machines. The first time Edwin had a whirl with the bulldozer, he had to use both hands to shift the clutch and gear handles and every time he turned the steering wheel it lifted his rear end out of the seat. Except for the time it went off the switchback, the little machine served well and Edwin spread tons of gravel with it on the mountain road. Thirty years later, he was working on this hill when a boulder about seven feet in diameter hit his Cat. It missed Edwin but broke the winch on the back of the machine. A few years after that, Edwin was working alone, grading the Hill, when the grader's motor quit. It lost both its steering and brakes, and tumbled over and down the bank, damaging itself and battering Edwin. He had a long climb up to the road and a long wait before anybody came along. He didn't give the mountain another chance.

Edwin was the only person on the Chilcotin/Bella Coola Road Crew to receive the provincial government's longtime service award, a gold watch. He had thirty-five years in February 1988. The Ministry of Highways privatized that same year, contracting all road maintenance to the private sector, and Edwin was too young to qualify for the early retirement package. He stayed on the job with the contractor, Caribou Road Services, until the grader incident in 1989.

# CHAPTER 21

# Breakups and Breakdowns

*Mired motorists muttered maledictions of roads and weather.*
—Indian Agent H. E. Taylor

The worst thing about breakup was the mud. There were many kinds of mud. Ordinary mud and sandy mud came in grey, brown and puke colour. Alkali mud was clingy and disgusting; swamp mud was scummy and looked like it was breeding something unspeakable; gumbo and loon shit—self-explanatory. There was also soupy mud, not too bad if it wasn't too deep, and slime. Slime was wet dust and vehicles didn't actually get stuck in it, they either slithered off it into the toolies or sat on it spinning their wheels. Soft spots and frost boils were something else again.

Rain always made the road slick and nurtured the low spots, but breakup was a bugger. Sensible people travelled at night, on the frost, as long as there was frost, but breakup complicated matters by coming at different times along the road. It started in lower Chilcotin and moved west. When it was springtime in Alexis Creek, it was breaking up at Tatla Lake. By the time the road was dry there, it was breaking up at Anahim, and the mountain road thawed later again. Some places, like Caribou Flats, were bad every year, but frost boils erupted anywhere and everywhere, and rivers and streams flooded whenever they pleased. The more the road was travelled in breakup, the worse it got. Trucks gouged big ruts and cars fell in them and high-centred. There was always a mess where anyone dug out of a hole. Cattle feet chewed up a soft road, and when it froze or dried, vehicles' wheels went up and down faster than pistons.

An experienced traveller carried an axe, jack, shovel and chains. Truckers carried chains for each wheel, cables, and

enough spare parts to rebuild the vehicle. Stan Dowling had a
logging jack that could lift his truck's front wheels two feet off
the ground as well as several ¾-inch cables.

The first trick was not to get stuck. In the snow, truckers used
single wheels. In spring, they drove in the ruts, through mud
holes instead of around them, and travelled on the frost. There
were a number of ways to free vehicles from mud. None were
easy; all required basic equipment, know-how and a little bit of
luck. If the vehicle bogged down in the timber, it could usually
be freed with the help of two cables and two trees. One end of
each cable was attached to its own tree, the other end to the dual
on the same side of the truck. With the engine going and the
truck in gear, the cables wound around the duals and the vehicle
pulled itself out of the hole.

Regular travellers were inventive. In the spring of 1940 Fred
Linder and Newt Clare, an Alexis Creek mechanic, spent three
days crossing Cahoose Flats near Anahim Lake. Cahoose Flats
was one long mud hole. The two men cut eight jackpine trees,
limbed them and tied them together with chunks of telephone
wire. They put one log bundle in front of the back wheels, the
other in the front, and spun the car into them. Then they put the
next bundles down, repeated the process, hauled the first two
bundles out, put them in front of the wheels and kept going until
they got out. Sam Colwell, who rode saddle horse from Kleena
Kleene to see why the telephone line was out, didn't say anything
when he saw what they'd done. He just replaced the line.

Stan Dowling said he was never stuck. "I always got out," he
pointed out. But sometimes it took a while. Breakup in 1940
lasted forever. The frost was deep in the ground and when the
road did start to dry, it rained. The bad spots stayed bad and new
ones grew. Phyllis Bryant was travelling home to Anahim with
Stan when he got into trouble in a deep gulley between two hills
past Towdystan. His front wheels were nicely up the incline when
the back wheels went down. The box was nearly touching the
ground so he couldn't back up. He unloaded the freight, and
found two poles. He put one pole lengthways under the truck,
with one end between the dual wheels. He tied a soft ¾-inch hay
cable to the duals through a hole in the wheel with a half hitch

on the hub, ran the cable along the pole, and tied it at the front end with a couple of half hitches. He put the other pole under the truck the same way on the other side. When he started the motor and released the clutch, the duals grabbed the poles and pulled them under, but they sank quite a way. He wondered if the motor would pull him out. He revved it a bit to test it, then revved to high speed and let out the clutch. It was close. The motor struggled, but the truck lifted out to dry land.

Stan unhooked the cables, reloaded his freight and got Mrs. Bryant home safely, but from then on the truck wouldn't go into high range. When he went to Vancouver with a load of furs some time later, he took it in to get it fixed. The mechanics couldn't get it apart and they couldn't get the axles out either, so they got a big sledgehammer and started pounding. The axle fell apart in a thousand pieces. Stan was glad that hadn't happened when he was in the mud hole. When he tried to winch his six wheeler, the Reo, out of a mud hole, one set of wheels twisted off. The housing wasn't heavy enough to stand the strain.

Most people pried themselves out of mud holes, and over the years the worst places collected poles and re-usable debris. When Jim Pomeroy took over as district engineer for Public Works, Peter Yells took him on a tour of Chilcotin. Caribou Flats had a three-hundred-yard-long mud hole and they got stuck in it. Peter found a pole for prying and another one for a fulcrum. Since Pomeroy was the guest, he got to sit on the end of the pole to hold the wheel out of the goosh while Peter poked sticks and branches under it. The mosquitoes were delighted with Pomeroy. He got so busy swatting them he lost his balance and fell off the pry pole. Down in the mud he went and down went the truck. The mud was mucousy. There was no way to wipe it off skin, never mind clothes. "That loon shit mud makes good mosquito repellent," Peter told him.

Roadside residents were used to helping mired or broken-down motorists, it went with the territory. According to Phil Robertson, "the only entertainment we had was watching some-one get out of trouble." Helping people was one thing, people helping themselves was another. There was a bad spot by Cold Springs Ranch at Riske Creek where the Jim Mackay family lived.

Mrs. Mackay finally appealed to Public Works for help. She said it was bad enough getting stuck in the hole herself, but everyone that went by got stuck there too, and they kept taking sections of her fence down to use for pry poles.

Breakdowns were as much a part of life as breakups. They could be counted on to happen as far away from help as possible. People did what they had to do to keep going. Phil and Joyce Robertson had an especially miserable time getting up Sheep Creek one winter day. Snow kept getting through the hood of their sedan delivery and melting on the distributor cap. Phil put the baby's rubber pants on the distributor and that worked. Other homemade remedies included making bearings out of bacon rind, pouring bleach on tires to give them traction on ice, and putting oatmeal or tobacco in radiators to stop leaks (oatmeal was a mess to clean out). Radiators always boiled dry where there was no water nearby so people used what they had. Beer often came to the rescue; it took nine bottles to fill most radiators. There are stories of men drinking the beer and peeing in the radiator, but no one owns up to doing it. If the only liquid available was whiskey, it went in the driver to sustain him while he waited or walked for help.

There have been conflicts between Chilcotin cattle and cars right from the beginning. In 1917, the Chilcotin stockman's association complained because motorists weren't paying attention to the signs warning them to watch for cattle on the road. R. J. Cotton suggested putting up signs telling the cows to beware of motorists and applying for an agriculturist to teach the cattle to read. At night, cattle that weren't road-wise would either freeze in headlights or charge. When an animal charged, its head would get the radiator. One fellow had a cow tumble off a bank into the back of his pickup. It put quite a dint in the cab.

Riske Creek rancher Mickey Martin's cattle managed to put Sheep Creek Bridge out completely in 1948. Martin, who frequently got himself into predicaments, was driving cattle to town and somehow, instead of the prescribed twenty-five head, a whole bunch of cattle got on the bridge at once and one of the cable anchors broke. Martin got the animals off safely but the bridge listed to one side so badly PWD closed it to traffic. Dowling happened to come to town before it was fixed, and the repair crew said he could go over empty, but couldn't go back loaded. Not wanting to make the long detour around Soda Creek and Meldrum Creek, he went back under the friendly cover of darkness when the crew wasn't there. The bridge was supposed to be good for only two tons, even with two cables, but he got across and it saved a long trip.

The friendly cover of darkness solved another problem for Chilcotin residents. As often as not they didn't get to Williams Lake in time to renew their vehicle licences, so they'd sneak in at night. The police never seemed to catch them.

Mickey Martin wasn't the only one to break Sheep Creek Bridge. The first time Shorty Fullerton took the PWD bulldozer, a TD20, across it, the needle beams broke. The story is that Fullerton, thinking the bridge was going, jumped off the machine and ran. When he looked back he saw it was still coming, so he got back on it and drove it off.

Horses often went to the aid of stuck vehicles, but Ole Nicola, the son of a Southern Carrier mother and a Scots father, pulled Wilfred Hodgson's truck out of the mud once with just himself. Nicola was incredibly strong. He took the saddle off his horse, rolled up the saddle blanket, tied a rope to each end of the blanket and tied the other ends to the front of the truck. He used the rolled blanket for a chest pad and pulled, and that was all the help Wilf needed.

No one ever agreed when the road was worse, at breakup or winter. You could shovel, or push or pull a vehicle out of snow. Mirey spots were something else. Nevertheless, winter travel had its moments. Ike Sing, who moved to Anahim Lake in December 1946 with his brother, drove a lengthened jeep rigged out as a camper. It was so cold he drove with a blanket over his knees. It took three days to get to Kleena Kleene from Williams Lake and then they had to turn back because the snow was so deep. The next day Fred Linder came along and he broke trail until his low gear snapped at Caribou Flats. Mrs. Andy Holte was with him, and the four of them stayed at Brink's range cabin. Wilf Hodgson came along the next day and led the parade until he broke an axle. He had a spare but they couldn't get the end of the broken axle out. They were near Fish Trap, and Ole was there. He whittled a jackpine stick for them to use to poke the axle end out. The little convoy arrived in Anahim Lake that night without further mishap.

Wilf Hodgson once unearthed an old musket when he was digging his truck out of a hole near Fish Trap. He wasn't in any humour at the time to appreciate what it might be. He just threw it in the back of the truck. By the time he remembered it, it had disappeared. He always wondered if it went back to the ambush of McDonald's packers in 1864.

The emergency cabin at Fish Trap was well used; it seemed to be in the right place for people in trouble. Roy Haines was stuck there for two days once when he was driving for Hodgsons. He had to unload all the freezables and his passenger, a dude in a suit and tie, didn't offer to help with anything. "Don't you wonder what I'm doing here?" the chap asked on the second day. Roy said he didn't really give a damn. The man said he'd been going to buy the outfit but he'd changed his mind.

Peter Yells and Roy's son Bob were snowploughing one winter near Fish Trap when the temperature dropped to -50°F. They were marooned for four days sharing the tiny cabin with Alf Lagler and two other men. They stoked the fire up until it was producing a major chinook, but it was so cold outside it didn't melt as much as a peephole in the frosted-up window. They ran out of food so they had to get going before it warmed up very much, but Lagler's truck wouldn't start.

"We can rig up a pipe and flame thrower and heat the truck and the battery," Bob suggested. It took some time to arrange this, but it worked, the motor started. As Peter and Bob walked to the Cat, Lagler bailed out of the truck yelling "Fire!" Lagler's truck was a discard from a construction company and it was worn out when he got it. Now smoke was belching out from behind the seat. Bob looked, and found some rags on fire from the blowtorch. He pulled them out and they all chucked snow behind the seat. The gas tank was there, and Bob thought it was leaking. "He's going to blow me up yet," Peter muttered, remembering the episode with the ether.

When Lagler went to go the second time, the rear end of the

truck was frozen and it wouldn't move. The bulldozer, an RD6, had ice growzers on it, and Bob tried nudging the truck. That didn't work so he backed up and took a run at it. That jarred the truck loose but it also caved in the back of it. Lagler didn't notice. The truck had no heater and one of the passengers kept scraping the ice off the inside of the windshield for Lagler to see. The two passengers were perishing cold but Lagler kept complaining how hot it was. When they stopped at Towdystan they discovered the seat was smouldering. Luckily it hadn't connected with the gas leak.

E. P. Lee put skis under the front wheels of his truck one winter but it didn't work. Harold Stuart used a shovel and it did. Harold's ski trip came about when he and George Telford were hauling a load of Telford's cattle to town. The road hadn't been ploughed but they were lurching along at a fair pace until they got to the A&P hill, about eight miles from town. Harold started to mutter and stopped the truck. "It won't steer," he said. A bearing had seized and the front wheel was locked. The cattle were thumping around in the box and the two stood in the swirling snow for a few minutes considering their options. It was a long walk to a phone. Then Harold climbed up to his overcab box and came down with a jack, pliers, some wire and his snow shovel. He jacked up the locked wheel and wired the flat blade of the shovel under it. "We'll ski in," he said, and they did, arriving safely at the stockyards half an hour later. George was impressed. Harold wasn't. "It wouldn't be worth a damn on gravel," he said.

Harold always had a trick or two up his sleeve. Dowling bought a pickup at an estate sale in Kamloops once, and the motor wouldn't idle. It stopped when the truck stopped and quit completely going downhill. Stan took it to a garage in Kamloops where a mechanic soaked the carburetor in a pail of cleaner, but that didn't help. Every time Stan took his foot off the gas, the motor stopped. When he reached Redstone he told Harold his troubles.

Harold crossed the wires by changing two wires, and told Stan to start the motor. After it backfired through the carburetor a few times, Harold replaced the wires to their original position, and the motor ran like a million dollars. There was no charge. Another time when the switch was giving Stan trouble, Harold wasn't home, but Christina told him to take the wires off and she gave him a clothespin to hold them together. "Just take the clothespin off when you want to stop," she said. That worked fine too.

When Stan sold his freight licence and mail contract to Hodgsons, he kept a five-ton 1953 Ford truck to haul cattle. The roads were nothing but washboard that fall, and vehicles took a terrible beating. On one trip the pounding pushed the back wheels on one side of the truck off the housing, and pushed the nuts off the housing, stripping the threads. When Stan realized something was amiss, he coasted to a stop and investigated. He found the wheels were out past the box. As it happened, the Purjue brothers, who had a garage at Alexis Creek, were travelling right behind him. They couldn't believe their eyes.

"It's a good thing you weren't going uphill or downhill and didn't use your brakes," they told Stan. They jacked up the box and pushed the wheels back on, burring the threads on the nuts so they wouldn't come off. The cattle were jumping around in the back of the truck all the while. Stan made it safely to town, but he decided he'd had enough close calls. It was time to leave trucking to someone else.

# CHAPTER 22

# A Bottle of Booze to a Tankful of Gas

*Many Chilcotin youths grew to manhood never knowing liquor could be mixed with anything, or sipped from a glass. They thought you opened the bottle, threw away the lid, and went at it.*
—Phil Robertson

There were plenty of seasoned drinkers in Chilcotin (teetotallers were a rare breed) and in Bella Coola too. A man was judged by his capacity to hold liquor. It was part of the mystique. The argument could be made that alcohol gave a rosier glow to a lifestyle that wasn't always wonderful. Chilcotin residents lived so far away from each other they didn't "get out" or even visit each other very often. When they did, they made the most of it. The Anahim Stampede was the one outing a year for some West Chilcotin families, and the idea there seemed to be to drink enough at one whack to last until the next one.

The on-road drinking was another matter entirely. It took courage to drive the road west, and courage came in bottles. Experienced travellers recommended a bottle of booze to every tankful of gas. "If you had enough to drink, the Bella Coola Hill looked like Becher's Prairie," trucker Don Widdis claimed. The road really did drive people to drink. It provided plenty of excuses—mirey spots, bumpy stretches, steep hills, sidling hills, slippery hills. The excuses had names—the A&P Hill, Sheep Creek Bridge, Sheep Creek Hill, Becher's Prairie, Hance's Timber, Lee's Hill, Tachadolier's, McClinchy Hill, Caribou Flats, the entire Bella Coola connection—all good reasons for imbibing nerve restorer.

Until the 1960s there was nowhere to get or buy a legal drink west of Alexis Creek, although there were home brew experts in every community, and bootleggers too. There was a licensed restaurant (beer and wine) at Anahim in the early sixties, and a liquor store in Bella Coola a few years later, but before that, booze was hard to come by. Those who could afford to bought a case of liquor at a time; everybody managed a bottle or two. Supplies didn't last long and whenever anyone went out, they stocked up for themselves and the neighbours too.

Most of the legal liquor came in on the freight trucks. Sometimes people phoned their order in, others had standing orders, and the more casual drinkers asked the driver to bring a bottle or two the next time he came. The drivers were supposed to have written orders for the liquor store but no one was ever challenged. Once in a while some of the liquor didn't reach its destination. The Hodgson drivers sometimes got thirsty, or they'd meet thirsty travellers. No one complained; they knew their order would show up the next week.

Then there was the illegal trade. Anyone with a mind to bootleg could make big money. There are wonderful stories of how people made fortunes at it because when a customer was dry enough he'd pay anything for a snort. One fellow from town was said to have put a low down payment on a new pickup truck, taken a load of liquor to an Anahim Stampede and returned three days later with enough cash to pay off the truck. When the law changed to allow status Indians to buy liquor, one woman thought she'd celebrate by buying the most expensive bottle of whiskey in the liquor store. She said she argued with the vendor when he told her the best bottle cost fifty dollars—people had been paying more than that for bootleg rotgut.

It was the custom of the country for motorists to stop and visit when they met. The westbound traffic provided the refreshment as those heading to town were unlikely to have anything but a thirst. Whiskey was the beverage of choice (west Chilcotin men liked Scotch), and a few opted for dark rum or beer. Mixers were unheard of, unless two liquors were mixed. One fellow served rye whiskey cut with creme de menthe as his Christmas cheer. Wine was known as goof, or porch climber, and was eyed with suspicion, but people heading to town after a dry spell were glad to get anything. People travelling together in different vehicles stopped frequently to compare notes over a nip, and most truckers carried stock with them. Don Widdis had a cooler with ice for his beer in hot weather. Travellers didn't get out of their vehicles to visit. There was very little traffic, drivers parked side by side and passed the jug from one window to the other.

The time it took to get from point A to point B certainly was another factor in the need for a bracer. As late as 1962, a Highways report said it took better than two hours to drive the 72 miles from Williams Lake to Alexis Creek, averaging 40 mph except for Sheep Creek, which took most of the time. From Alexis Creek to Kleena Kleene, 81 miles, average speed 30 mph. From Kleena Kleene to Bella Coola, 155 miles, recommended speed 25 mph, except for the "spectacular 14-mile long hill" which slowed everyone down. The report was talking about good weather and didn't tell all the story. Another hindrance to speedy travel were the P&P (puke and pee) stops, inevitable and innumerable if there were children passengers. It was a given that everyone in a vehicle who got carsick, or had to go, required a separate stop for each event.

It was possible to make better time in the winter when the potholes and washboard filled in with snow or ice, but then the road was slippy and speeders were apt to end up with their nose stuck in a snowbank, and it took a lot of shovelling to get out. By spring, the snowbanks told lots of stories. When top speed was ten miles an hour, a sip or two helped pass the time.

Another argument for carrying a crock was that if you did get stuck, you had something to sustain yourself until help came along. One trucker was hauling a piano and two passengers from Bella Coola when they got stuck, but they had some refreshment with them, so they got in the back and one of the passengers played the piano. They were singing and having a great time when the road crew arrived to rescue them. It certainly beat sitting in a crowded cab all night. Sometimes, especially in winter, local men took days to reach their destination because they waylaid themselves along the way. They'd stop somewhere to visit a friend, and if either party had some booze they'd stay until it was gone. Often the drivers were soused when they stopped for a visit and they stayed until they sobered up. Riske Creek was a popular stopping place for traffic coming and going. Men would "stop over" at Scotts or the Lodge. Dru Hodgson had two locals stay at the lodge one Christmas Eve. They were headed home but she couldn't get them moving even though one of them kept muttering about getting "cold turkey" when he got home. The police stopped Fred Linder and Bill Woods once at Kleena Kleene when the two were on their way home from an Anahim Stampede. They'd stopped to visit here and there and Bill, who was driving, was slightly the worse for wear. The officer told him to drive slowly. Bill said he didn't see how he could drive any slower than he had been, it had taken him three days to get from Anahim.

Some men were noted for their capacity for drink. E. P. Lee drank a bottle of brandy a day in his later years. He said the government watered it down so much it hardly counted. When told Don Widdis wasn't drinking any more, one old-timer commented, "he don't drive that road no more either."

Drinking drivers rarely damaged anyone but themselves—there was rarely anyone else on the road to damage—but the Drummer certainly kept an eye out. There was the odd accident. Some cowboys never did get the hang of driving. Grover Hance wasn't the best driver in the world at the best of times. He could be a hazard when he'd had a few. Tommy Hodgson was stopped at Chimney Creek once putting water in his radiator when he saw Grover coming down the road in his Starr car. The Hodgson truck was well off the road but Grover drove smack into it. Things like that happened sometimes.

Bella Coola residents thought their road was the worst of all but few of them drove the Chilcotin Road in its prime. The Bella Coola Hill truly was the granddad of all highway horrors, but Sheep Creek Bridge and Hill came as a set and there was no time for motorists to get their wits about them between one and the other. On the other hand, the Bella Coola traffic had all three to contend with. Oddly enough, the people who got in the most trouble were cold sober at the time. Perhaps it did help to be squiffed.

Every hill claimed victims. One of the first pickup trucks to come out of Bella Coola navigated all the hazards safely until it reached McClinchy Hill. There it barrelled off the edge and landed in a tree. No one rescued it and it hung there for years. Lee's hill caused numerous accidents, some of them fatal. People did go off and over Sheep Creek Hill. Riske Creek road foreman Dave Chesney earned his place in Chilcotin folklore when he put his Pontiac car in reverse instead of low going around Cape Horn and it flew off backwards. When Ed and Fanny Boyd of Chezacut and another couple went off Cape Horn, Mrs. Boyd suffered the only injury—she crawled over a dead porcupine on her scramble back up to the road.

Sheep Creek terrorized three generations. The Bella Coola Hill is working on its second. The original ledge blasted up the rock face was wider than it looked, but not much. There were pullouts for passing, and room at the hairpin turns to stop and catch one's breath, have a drink, or cry while the brakes or engine cooled. Anything longer than a pickup truck had to back up several times to get around the corners. If vehicles met between the pullouts (and they always did), there was no rule as to who should back up. The one going up with the boiling radiator usually gave way to the one going down with the smoking brakes, but the driver containing the most booze was expected to do the honours. It was a special thrill to go down the hill in a car with an automatic transmission. No one ever agreed which was worse, going up or down. The descending driver has the inside track twice (there are some who swear there is no inside track). Either way, both sides of the road were crumbly—the outside edge crumbling down over the bank, the inside edge sloughing onto the road. The road was so narrow in places the person on the outside edge couldn't see any road underneath. In the winter, the sun hit the side of the mountain and ice built up on the inside, tilting the road to the drop-off side. Faced with this situation, Bella Coola resident Norman Saugstad told his terrified wife it was all right to cry, but she couldn't get out and walk because it was too slippery.

Going down, passengers would get such a grip on the door handle or dash they almost had to be pried loose at the bottom. Their legs ached from trying to put on the brakes. Nothing matches the terror of being in a vehicle slipping backwards. The first grade at the bottom of the hill was covered with little pebbles.

Wheels spun out on them going up—it could take a few runs to get up the pitch. Going down, a vehicle with too much speed could skate right off the road on them.

Morton Casperson was the first to tumble off the Hill. A gamey old gaffer from Bella Coola, Morton had a meadow near Anahim. He was going up the hill with a team and wagon when his horses gave out. They couldn't pull the load and they couldn't hold it either. Morton jumped off to put rocks under the wheels, but the wagon rolled before he could do it. It went over the hill, pulling the team with it. There were chickens on board and they got loose, they scattered all over the mountain. The horses survived. Thirty years later a policeman went over in the dark. The vehicle slithered down the bank, stopping on an outcropping of rock. When the driver got out, there was nothing to stand on, and he slithered down farther. He was bumped, bruised and shocked. Some joker marked the spot with a sign that said "Taylor's Way." It had an arrow pointing straight down. Others went off the hill, but for forty years a rock or a tree or a snowbank always got in the way to prevent disaster. Some say there were so few serious accidents because people drove scared.

In the 1960s, a passenger/freight service between Bella Coola and Williams Lake helped keep the liquor supply going. The owner was a bit before his time. The road wasn't ready for him—or he wasn't ready for the road. He didn't follow the rules. He never checked the state of the Bella Coola Road at either end before he sallied forth. Once he drove right around a barricade and a big "Road Closed" sign. He also drove too fast. In winter, he either went off the road himself or forced someone else off at least once every trip. The Anahim crew spent a lot of time rescuing him from what they considered to be avoidable trouble, and he wasn't their favourite motorist. The final insult came when he schmucked into Anahim foreman Bob French one winter day on the mountain. The road got narrower as the snowbanks were ploughed higher, and by February most of the road was one-way traffic. It behooved travellers to drive cautiously because they were just as likely to meet a moose as another vehicle, and moose were apt to panic and charge the oncoming vehicle when they couldn't get over the snowbank. This particular day there was a clutch of Highways officials on the road behind the bus on their way to Anahim. French saw the bus coming and headed into the snowbank, but it got him on the door. French had been in town the day before, and he'd brought a bottle of whiskey back with him—an absolute no-no in a government vehicle. He'd forgotten to take it out. It was stashed behind the seat and that's just where the bus hit. He was still trapped in the cab when the highway officials came along, and he could smell the alcohol. Fortunately it was so cold outside no one else could. He decided the close call was a warning about breaking rules, but he was peeved that he'd lost his bottle. A few weeks later when the bus went by, the crew could hear a loose tire chain whacking the bottom of the bus. The driver didn't stop so they couldn't tell him. On their way home they found a box of booze—obviously someone's liquor order—in the middle of the road. It must have fallen through the hole made by the chains. Three of the bottles weren't broken so the crew kept them. They figured they'd earned them.

The road drinking was at its peak in the late 1940s and 50s. As civilization pushed west, the on-road hospitality disappeared in lower Chilcotin. Even the thirstiest traveller didn't care to stop anywhere near the road in the dust of the 1960s, although travellers still stopped at the lodges or wherever to visit. It was a different matter beyond Puntzi and for a long, long time the Bella Coola Road encouraged drinking. If all the beer, wine and whiskey bottles thrown out of vehicles over the years had been collected, they would have provided enough fill for every low spot.

# CHAPTER 23

# The Winds of Change

*We have six trucks and one passenger vehicle on the road at the
present time. We have had so few accidents over the years that our
insurance company lowered our rates, but now we have had six
collisions in six weeks. This road is inadequate for logging trucks.*
—Joe Gillis, Hodgson Brothers, 1961

The winds of change began blowing over Chilcotin,
gently at first, in the 1950s. Some came from the east, across the
Fraser River, with the lumbermen. Some came from the south—
high land prices and troubled times in the United States brought
newcomers to the country. Some winds blew because times
change, even in Chilcotin, even for the Drummer.

When the lumbermen discovered Chilcotin in the 1950s, they
brought little mills that popped up like mushrooms after a rain.
At one point there were twenty-five of them strung out between
Riske Creek and Tatlayoko. The logging didn't reach beyond
Tatla Lake—nobody wanted the scrubby jackpines—but they
brought well-paying jobs to the rest of Chilcotin. The only people
who didn't like that were the ranchers. They couldn't match the
wages (neither could Public Works). Some mill jobs paid more
an hour than ranchers paid a day in hay time. When the hay crews
deserted for mill jobs, ranchers had to replace them with machines, which had a way of breeding. Tractors needed something
to pull, and when there were too many things to pull, the rancher
needed another tractor. One thing led to another.

Communities developed as families moved closer to the mills.
With the Puntzi air base and the mills, the late 1950s were good
times for the Chilcotin valley and Riske Creek area service indus-
tries. People started cafes, garages and motels all along the road.
After the demise of Becher House, Bert Roberts' place was the
"city centre" at Riske Creek. He sold the store, cafe and cabins

to a Vancouver couple, Joan and Bob Scott, just as the logging boom hit the area. Joan was English, city born and bred; she was just getting used to coping with country life and Chilcotin people when the first logging shows hit the Riske Creek area. Truckers left Williams Lake before anything was open, and they stopped at Scott's to fuel up their rigs and themselves. When Joan got up at 5:00 in the morning, she usually found three or four fellows napping in her living room, waiting for the gas pumps and breakfast. They were back for lunch or dinner, depending on the length of their run.

When the two big companies, Lignum's and P&T, decided to handle everything at their large mills in Williams Lake, they closed their Chilcotin operations. The little mills disappeared as quickly as they came. The community that developed near Lignum's Mill at Tatlayoko disappeared, leaving a new three-room school and three modern teacherages. The Chilanko Forks area slowed down when the P&T Mill left and the Puntzi air base closed. Some of the roadside businesses disappeared, but most survived. Red Allison bought the Scott place in the late sixties, about the same time Wilf and Dru Hodgson took over the Riske Creek Lodge. Gordon and Betty Jasper's place is now the Big B. The cafe and motel at Lee's Corner became permanent fixtures, but Alexis Creek fared the best of all. It kept the additional forest service personnel, the second service station, and the motel and restaurant. What was even better, Hydro power reached the community—no more light plants.

The logging didn't stop in Chilcotin. Trucks kept hauling Chilcotin logs to the mills in Williams Lake and the winds they created changed Chilcotin most of all. The road between Williams Lake and Puntzi was a treat when it was first rebuilt for the air base. People were just getting used to it when the logging trucks hit the road. Most Chilcotin ranchers were hauling cattle by the early 1950s, but cattle trucks, like the freight trucks, were gentle with the road. Logging trucks were not. They were big, and in a hurry. They had no time for anybody or anything. They were a shock to Chilcotin where people and their vehicles were used to low gear. The dirt road was no match for the pounding it took from the big trucks. It was hard to say which was worse, the gut-wrenching potholes, rocks and washboard, or the dust. Even the smallest, slowest car left a memory of dust floating over the fields and through the forests. The logging trucks spewed out great clouds of it, brown, rolling, choking dust that blinded everything coming behind them or meeting them. It hid the road for miles. It smarted eyes, plugged nostrils, gritted up teeth and turned hair grey and stiff. People who wore glasses had to stop every few miles and clean the dust off the inside of the lenses. A crust of dust hung on the trees and grass, and coated the hay in roadside fields. There was no way to keep it out of vehicles and motorists had a choice of choking on the old dust roiling around inside, or opening the windows and choking on fresh stuff billowing in. There was no way to keep it out of buildings. It coated clothes and food, and people who lived beside the road nearly went mad with it. Highways crews put chemicals and oil on the sections through settlements, but some people were allergic to the chemicals, and the oil was mucky. The road crews couldn't keep up with it anyway. People prayed for rain.

And then there were rocks. The lower road was down to bedrock. Where it wasn't washboard, it was cobbled. Rocks skulked in the dust to sabotage oil pans, bend tie rod ends, and poke holes in tires. Distances were measured by the number of flat tires. Truck wheels pegged stones at windshields with deadly accuracy and kicked up gravel to pockmark paint jobs. People traded their vehicles after two years of replacing shock absorbers, punctured gas tanks and broken speedometer cables. Everything on a vehicle rattled after a few months on the road and car dealers weren't crazy about taking Chilcotin trade-ins.

Residents had innovative ideas to cope with the rocks. Harold Stuart suggested pounding them down with a pile driver. Someone else thought hovercraft might be the answer. Beleaguered travellers took their frustrations out on the highways crew, who weren't any happier about the road than anyone else. They drove it, worked it, wore it, breathed it and ate it all day long.

In winter, the road got narrower as the snowbanks got higher. With no dust to warn them of oncoming traffic, vehicles that met on a corner had the choice of hitting head-on or whamming into the snowbank, which could be as hard as concrete. The government ignored the situation. Even with some truckers making several trips daily, the traffic count over Sheep Creek bridge in 1961 was only 220 vehicles a day. Traffic petered out beyond Tatla Lake and the Bella Coola connection might not see a vehicle for days.

The 1960s brought another change to Chilcotin—new settlers. Like those who came before them, the newcomers were seeking new frontiers, freedom, or both. Some tried to live the old way, off the land, not realizing that never had worked very well in Chilcotin, where the growing season is limited by frosts that come most months of the year. The new ranchers who came brought new ways with them. They expected their property to have surveyed boundaries and secured water rights—new ideas to west Chilcotin. The newcomers also expected something in the way of service. As more than one old-timer complained, "these people come here to get away from it all and then they want to bring it all here with them." The "it" included better schools, better mail and telephone services, better roads—all the trappings of civilization.

West Chilcotin was relatively untouched by the logging activity. When the Escotts and Baxters bought Ike Sing's place at Anahim Lake in 1960, the downtown population was nine. A few Ulkatcho families lived full-time on the reserve next door where there was a school, a Catholic church and a resident priest. The school principal was Missionary Sister of Christ the King, Sister St. Paul, who had moved west from Anaham Reserve. St. Paul built the school herself (she said if you could sew you could carpenter), and as she was a registered nurse as well as a teacher, she provided emergency health care to the whole community.

Don and Marilyn Baxter were among the newcomers who heard—and challenged—the Drummer. While Don "learned his furs" and hunted for a decent water supply, Marilyn looked after the cabins and restaurant, minded the kids and learned the woman's version of the Chilcotin Two-Step—running flat out so as not to lose ground. In 1969 Baxters built a modern motel and restaurant, and sold the store. Like Dowling, Baxter was a big-city import, and like Dowling, he gave civilization a few hard shoves in the direction of Anahim. Over the years, he was a major mover

*Betty's store, Tatla Lake. Bob Graham's first store, which opened in the early 1930s, was a small log cabin dating back to Benny Franklin's days at the turn of the century. After daughter Betty took over the duties of storekeeper and postmistress, the Grahams built this new store. Today this building is a private residence.*

*A. C. Christensen Ltd., Anahim Lake. Norwegian colonist Adolph Christensen took up storekeeping after he arrived in Bella Coola in 1898. In the 1930s, his son Andy moved to Anahim Lake to ranch and opened a "branch plant" there. He packed all their supplies for the store and the ranch up the mountain from Bella Coola, a three-day trip one way. When Andy and his wife Dorothy retired, their son Darcy took over. When the old store burned, he replaced it with this new building but he carries the same stock–some of everything.*

*Alexis Creek post office. Tommy Lee sorting the mail at his store in Alexis Creek. At best the mail arrived twice a week, at which time everyone in the neighbourhood converged on the store to collect their post.*

*Lasting landmark. Not much of Tachadolier's log cabin is left, but it is still a Chilcotin landmark, located across the road from Tachadolier's meadow and always spelled wrong on road maps.*

and shaker in getting a new integrated school, hydro and an airport. He organized lobby groups, hassled for road improvements and served as school trustee and regional district representative. Marilyn won Lester Dorsey's highest accolade. "Doesn't matter when you get there, or what you do when you do get there, she's always nice about it," he said.

All West Chilcotin residents went to "town" occasionally, the newcomers more often than most, and they too began to notice the deterioration of the road. Tom Chignell was still smarting over the Chilanko Diversion when he asked Public Works to survey the road. "I grant a survey would have to be made from a helicopter or the poor engineer would be torn apart by the different factions, but an impartial survey is absolutely essential before any more taxpayers' dollars are spent," he wrote. "Contracts should be let for work where presently storekeepers with a bulldozer can build a road wherever they like with antiquated machines. Vested interests dictate where roads go and how many times they are built and rebuilt. They always end up in the wrong place." District Engineer Herb Coupe agreed. In 1960, after due process (Pete Yells asked everyone he thought might be interested), PWD named the Bella Coola/Chilcotin Road, and Coupe tried to convince Victoria that a surveyed road would be less costly in the long run.

Road improvements were not in the stars, but Highways Minister Gaglardi did replace Sheep Creek Bridge. He had to. The Bridge was mended and reinforced many times over the years, but it was too old to be bothered with new responsibilities. It always was noisy, but now it complained bitterly when anything went near it. It sagged so much in the middle when loaded trucks went across that they actually went uphill as they drove off. The new bridge is a spiffy two-lane concrete-and-steel structure. Called the Fraser River Bridge, it opened for traffic in September 1962, fifty-eight years to the month after the first bridge opened.

Highways officials destroyed the old bridge, afraid people would get into trouble on it. Experts were imported to blow it to smithereens so the debris wouldn't cause problems in the Fraser River. The blasting crew put powder in every nook and cranny. District superintendent Frank Blunden and Peter Yells were on hand to watch the demolition. When everything was ready, the fuses were lit. There was an impressive whoompf. The bridge shook itself, ruffled its feathers a bit, and settled back down in place as though nothing had happened. The powder people had to start all over again. The second time, the east side of the bridge obediently blew to bits but a large chunk of the centre section lifted in the air, hung there for a second, then plopped into the river. Onlookers watched in wonder as it floated grandly downstream and out of sight, guard rails and all. No one ever heard what became of it. The blast splintered the west tower, but it wouldn't fall. Blunden's crew finally burned it. The stone piers on the east side are still standing.

Once the new bridge was in use, the highways ministry was bombarded by letters and petitions from truckers, loggers and ranchers who said the bridge was useless until something was done about the Hill. Trucks couldn't get around the hairpin turns. The grade was too steep for the heavy loads. The roadway was so narrow big trucks had to hug the bank to make the turns, and they kept clonking gobs of dirt and clay on the road. When it rained, the surface was slithery. There were numerous accidents. Lige Gurr was senior foreman in Williams Lake, and he must have been tired of trying to keep hills in working order.

Surveys, plans and estimates were prepared for Hill work, but no money was forthcoming. The Unemployment Insurance Commission got into the act, saying jobs were needed. Ranchers shouted into MLA Bill Speare's ear, complaining their cattle got to market dehydrated and shrunken because of the road. The Cariboo Lumber Manufacturers' Association, Chilcotin businessmen and residents campaigned to "Get the Hill Out of the Road." About this time Bella Coola residents started getting testy too. "The ultimate destiny of Bella Coola certainly includes a highway outlet for Prince George," the Board of Trade harumphed in 1964. Businessmen from Ocean Falls, Bella Coola, Chilcotin and Williams Lake formed the Bella Coola/Chilcotin Road Association and joined the fray. The latter group were convinced Highways Minister Phil Gaglardi had $20 million stashed away somewhere.

In June 1964, the assorted road lobby groups got together to stage a major rally at Anahim Lake. It was Highways practice, if not actual policy, to tart up the roads whenever any government dignitary was due to visit. Word of the impending visit would come from above, and crews were hauled off whatever they were doing to grade and patch so the visitor would think the road was in great shape. This irked residents (and road crews) who thought the people who controlled the money should see the road as it truly was.

The rally was planned for June because the organizers knew the Bella Coola road wouldn't have recovered from breakup yet and the Chilcotin road still would be acned with frost boils. The west end road foremen, Mack Gurr and Bob French, were not delighted about the event. They knew what cavalcades of motorists would do to the soft road. In the days before the rally they hoped for hot sun and drying winds, but they didn't get it. As the big day approached they dutifully pulled out all the stops to have their sections passable, if not presentable, for the expected horde of protesters. The Anahim and Bella Coola crews were on the job—and overtime—on the meeting day, a Saturday. Gaglardi was to fly in to One Eye Lake where Blunden would meet him and drive on to Anahim. Mrs. French was home manning the base radio set in case of emergencies. About the time Gaglardi's plane was due, the Bella Coola crew decided to visit on their mobile radios. Normally no one could hear them, but if Gaglardi happened to be listening from the plane he could, and Mrs. French thought some of their comments inappropriate for ministerial ears. She leaned on the base set's transmitter, effectively drowning them out.

In the meantime, several hundred people were waiting at the community hall on the Stampede grounds. The Anahim women had been busy making stew, while the Bella Coola women barbecued salmon. Representatives from the different groups had speeches prepared and everyone was ready to lambast the highways minister. They waited and waited.

The plane landed at One Eye on time, but with retired deputy minister Evan Jones aboard instead of Gaglardi. It was a smooth move: Jones was well liked and no one would give him any static. They knew he couldn't do anything anyway. Blunden was waiting, and they set off for Anahim. Barrelling along Caribou Flats, Blunden blopped into the only frost boil on the road between Williams Lake and Anahim. He landed to stay. He radioed for help, but couldn't get through to anyone because the busybody on the base set had everything blatted out.

When Gaglardi didn't show up, a helicopter was sent to investigate (not everybody braved the road to get to the meeting). When the pilot found the official party stuck in the frost boil, he flew Jones to the meeting, and let Bob French know to rescue Blunden. The event made the radio news and there were pictures of the bogged vehicle in the newspaper. Jones thoroughly enjoyed the adventure. Everyone had a field day—except Blunden. He blamed the lack of radio reception on a dead spot. Mrs. French didn't enlighten him.

It was probably a coincidence that tenders for reconstructing Sheep Creek were called later that year. Known officially as the Sword Creek project, the estimated cost of rebuilding the 4.88 miles was $668,000. Ben Ginter Construction from Prince George was the lowest bidder. The project was plagued by problems. Like the Bridge, the Hill didn't make things easy for anyone. There were constant revisions. Drainage wasn't included in the original contract, and Sword Creek flooded, wrecking the road and rancher Roy Thompson's alfalfa field. Ginter's crew camped in the fields and didn't pay. Fences were knocked down, letting Thompson's bulls get in with the heifers, and the calves arrived in January. Thompson said his water licence was destroyed by the project. Highways built a new intake and weir but there was no water in it because the springs were under the new road. Ginter had problems too. Inspectors found faults and he didn't get everything fixed to their satisfaction until 1968. Travellers were happy with the new hill, especially the truckers, but nothing was done with the rest of the road. It just kept getting worse. "It is a potholed, narrow, blind-cornered, frost-boiled, washboard cow path," a citizens' group complained to MLA Bill Speare. "The standard of maintenance on the Chilcotin road is not poor. It's nonexistent," Blunden replied. Nobody, except the people who used it, seemed to care.

In the spring of 1967, the Bella Coola Board of Trade sprang into action again with a series of advertisements in the *Vancouver Sun*. The ads began "This is the neglected land" and went on to say why. The Board hoped to embarrass the government into providing an all-weather road, but it didn't work. The government did spend money in Bella Coola, but it usually went to repair flood damage. One flood in the fall of 1965 destroyed three major bridges and five miles of road in the valley. They had to put a ferry at Hagensborg, where coho salmon were actually swimming up the highway. Phil Gaglardi announced he was heartened by the employees' exemplary spirit in working under all conditions—he probably didn't realize that's how they always worked. It cost $286,500 for temporary repairs.

On April 1, 1966, the North Vancouver Highways District turned Bella Coola over to Williams Lake. It made sense from an administrative point of view but Bella Coola was poorer for it. Williams Lake never had enough money. Crews improvised or did without. Blunden was such a good scrounger one of the senior officials called him a bloody menace. His 1964 budget included $22,000 for the district bridge crew, enough to pay the crew's wages, but not a penny for a plank or nail. Blunden had to think of something so he had his crew do the brushing on the Sheep Creek Hill project, and he made a deal with Lignum's mill to trade the usable logs for cut lumber for the bridge crew. Some of his other ideas weren't so good. He and Bill Telford thought they'd save a dollar by using old grader blades for mileage markers, but when someone went off the road and hit one, it sliced the car.

In 1968 W. D. Black became highways minister, and he too neglected Chilcotin. The traffic counts no doubt had something to do with the lack of activity. In 1965, 265 vehicles a day crossed the Fraser River Bridge (a gain of five in four years); in 1966, 265 vehicles; in 1967, 270; in 1968, 300. The figures were for the east end, the busiest part of the road. "The Chilcotin is early ranching country. The access road carries little traffic over great distances to lightly populated areas. We are improving it as quickly as we consider warranted for the limited use," Black wrote in his 1970 annual report.

As the road got worse the complaints got louder. Tourists

complained; ranchers complained; truckers complained. Don Baxter wanted to put a sign at the top of Sheep Creek saying "This is God's country, don't drive like hell," but the authorities wouldn't go for it. Elected officials complained. "About halfway between Williams Lake and Bella Coola are two signs warning people of bumps on the road," Skeena MP Frank Howard wrote in a letter to the Williams Lake *Tribune*. "The signs should be reversed to state on the one 'Bump all the way to Bella Coola' and on the other 'Bump all the way to Williams Lake.'" Alex Fraser, the mayor of Quesnel and a regular at Anahim Lake stampedes, complained too. He kept complaining after he was elected Cariboo MLA in 1969, but nothing much happened. "The snowbanks are five feet high in places," he wrote in 1970. "The road is so narrow there is nowhere for traffic to go but into the front end of the oncoming vehicles or over the bank."

About this time Blunden was transferred to sunnier climes—and better roads—and Jim Steven, a quiet MOH veteran with a slightly off-the-wall sense of humour, fell heir to the Williams Lake District and the Bella Coola/Chilcotin Road. He needed the sense of humour. During the 1960s and 1970s, Highway 20 may not have been the worst road in Canada, it may not even have been the worst road in the province, but its reputation as the longest worst road anywhere was never challenged. Steven spent almost as much time coping with complaints about Highway 20 as he did planning road work—the government simply wasn't prepared to spend any money on it. Once he received a letter from a tourist who actually had some kind words to say about the road crew. Steven replied that he so seldom heard anything but complaints, he really didn't know how to respond to a bouquet.

While the road was the main topic of concern in Chilcotin and Bella Coola (Cliff Kopas said he didn't know what people would talk about if it wasn't for the road) valley residents faced another problem in the 1960s. The land taken up by the colonists was properly surveyed, but as time went by, the original owners subdivided. The owners of the subdivisions subdivided. The buyers and sellers measured off the property, drew up agreements and had the local Notary Public do the paperwork. There was no problem until the Social Credit government brought in the

Homeowner's grant, and to get it, property owners had to show proof of ownership. That's when many Bella Coola landowners discovered they had no legal title. None of the land transactions had been registered in Victoria.

Cliff Kopas talked to a lawyer who contacted Vic Cockford, a Vancouver surveyor. Cockford looked at a map, then agreed to go to Bella Coola if ten people would have their land surveyed. This was arranged, and Cockford made his first trip in 1962. He spent the next twenty years sorting out the muddle. He tromped through the thick, dank underbrush, surveying properties from spring to snowfall. He was damp and muddy much of the time. Once he had the groundwork done, he had to deal with the massive amounts of paperwork required to register the land. It was a bureaucratic nightmare of magnificent proportions. The lands office in Victoria didn't even know where Bella Coola was until Cockford told them. Each application had to be sent or taken to Highways approving officers, first in North Vancouver, later in Williams Lake. Nothing fit the rules. Valley residents didn't understand about rules anyway, they'd leave notes on trees asking Cockford to do this or that. By the time he had all the subdivisions on paper, he knew the valley very, very well, almost every inch of it. He also knew all about Bella Coola's communications and transportation problems.

The valley road wasn't surveyed either, of course. It went where it pleased with stunning disregard for legal right-of-way or land ownership. When Cockford discovered the road crew had moved some survey posts (probably during a Minor Betterment), the matter found its way to the provincial Legislature. Frank Clapp, Highways' chief surveyor, was sent to investigate. He was startled by the extent of the chaos and set about gazetting the roads in the valley, another lengthy process. However, it did convince someone in government that perhaps it was time to look at the survey situation for the entire Bella Coola/Chilcotin Road. As late as 1982, little of Highway 20 was surveyed between Redstone and Nutsatsum.

# CHAPTER 24

# Highballing and Eyeballing to the Pacific

*We could go on forever dumping money and gravel and time into the Chilcotin Road, but I think the point is to upgrade most of it to a certain standard, then maintain it at that standard until it can be blacktopped.*
—A. V. Fraser, Cariboo MLA and Minister of Highways, 1976

In Victoria, he was called the King of the Cariboo. In the Cariboo he was called Alex. His name was Alexander Vaughn Fraser and he was the Cariboo MLA from 1969 until his death in 1989. He was the minister in charge of highways from 1975 to 1987. During his reign he did more to improve the Chilcotin Road than anyone else ever did before (even if you put what they did all together) and probably more than anyone ever will again.

Alex was born and raised in Quesnel and he was no stranger to politics. His father, John A. Fraser, was the Cariboo MLA from 1912 to 1916 and the Cariboo MP from 1925 to 1935. Alex himself was mayor of Quesnel for twenty years before plunging into provincial politics. He was no stranger to BC's roads either: he was a partner in the first freight line to provide a regular service from Quesnel to Vancouver, and he spent years on the road behind the wheel of a freight truck. Alex was a staunch member of the Social Credit party, and during the New Democratic Party's little burp in power between 1972 and 1975, he made a name for himself by constantly giving them hell in the legislature and out. When the Socreds returned to power in 1975, Alex became Highways and Transportation Minister and put his rowdy days behind him. Over the next dozen years he earned the respect of friend and foe in the Cariboo, and the absolute devotion of Chilcotin residents.

*Mr. Minister. More attention was paid to the Chilcotin Road during Alex Fraser's ten years as highways minister than ever before. Cariboo born and bred, Fraser was a trucker before he went into politics. He was a regular at the Anahim Lake stampedes, who knew firsthand the horrors of the Chilcotin Road. When he was mayor of Quesnel, he was known to let his hair down at stampedes, but as highways minister he was the polished politician, although he wore his cowboy hat on appropriate occasions.*

*Chilcotin lifeline. Residents were sad when the original Sheep Creek Bridge linking Chilcotin to the rest of the Cariboo was demolished, but they weren't sorry to see the new one built. The old span, built in 1904, squeaked and squawked and complained under the lightest load; those crossing with heavy loads said it felt like driving on waves. It also sagged in the middle so that trucks drove up uphill. The new bridge opened in 1962. Safe and handsome, it does its job quietly and efficiently, but it doesn't have much character. Nobody has any stories to tell about it.*

*Sheep Creek Hill, 1960s. Elijah Gurr on the road, during construction.*

Chilcotin ranchers lean pretty far to the right politically—they voted Social Credit anyway—but their support for Fraser was magnificent. He was personally acquainted with the Chilcotin Road, as he was a regular at the Anahim Stampede. As minister, he never missed the stampede or any other major Chilcotin gathering, be it the Alexis Creek Pioneers Day or an old-timer's funeral. He had a knack of remembering people's names and circumstances and when asked for help he always responded, even if the help wasn't forthcoming. His personal touch was his trademark.

Contrary to common belief, the NDP didn't do "nothing" for Highway 20, but their most significant contribution to the provincial highways system was the establishment of maintenance management. This program used both manual and computerized performance reports which compared actual work with planned work to assist with planning, scheduling and controlling Highways activities. A "performance" budget allocated manpower, equipment and materials for work based on set standards, department policy and available funds. The system was meant to curtail political interference in road work, if not to eliminate it entirely, but no system devised by man could have kept Alex Fraser from finding ways to please his constituents.

Shortly after his appointment to cabinet, Alex vowed to pave the road to Alexis Creek. No plan or budget had ever been set up for rebuilding the Chilcotin Road because previous ministries hadn't even considered it. Longtime residents feared pavement would be the death of Chilcotin, making it more accessible to polluters, vandals, speedsters, tourists and other undesirable elements. (By 1977 the conversion to metric signs was completed and the old-timers didn't like that either.) There were numerous legal and political hassles during reconstruction because the new road ventured where none had gone before. Eventually, however, the deed was done.

Although Alex's heart (and voters) were in Chilcotin, he didn't ignore Bella Coola. He knew most people felt the Bella Coola connection was in the wrong place. The question was whether to leave it or change it. Alex investigated the possibility of rerouting the road over the Lunaas Trail. He had surveys and studies done, but there were two major problems. One was political: the Towdystan route bypassed Anahim and Nimpo Lakes, which didn't sit well with the businessmen in those communities. Second, there were too many environmental concerns to build a road through the Hotnarko and Atnarko river valleys. After surveying the alternatives, Fraser opted to upgrade the sections of the existing road that were in the most lamentable condition. He had the switchbacks on the Bella Coola Hill widened to accommodate trucks with trailers. Before that, truckers had to leave the pup at the top of the hill and come back to unload it—a 110-mile round trip. He also spent $2 million rebuilding the road between Alexis Creek and Anahim Lake, hiring local equipment on a day-labour basis. The road work gave a tremendous boost to the west Chilcotin economy, and even the balkiest old-timer found the new road a joy to travel on after years of bumpy dirt.

In 1978 the supply boat, the *Northern Prince*, stopped calling at Bella Coola. The next year, the bottom fell out of the road west of Anahim. It hadn't rained much that spring but there were lots of frost boils because of frost penetration. Alex said the freight trucks had to get through regardless of cost. Bella Coola foreman Bill Thatcher and Anahim foreman Doug McKee had three crews working around the clock for three weeks laying jackpine corduroy (just like the old days) to support fill. They made a complete new roadway between Morton's Meadow and Green River. They used ten-foot poles which shot up when the freight trucks ran over them, but they didn't have to do much fill, the mud squished up and covered the poles. That same year the roads passing through Riske Creek, Alexis Creek and Anahim Lake were paved instead of oiled, making life cleaner and more pleasant for the merchants.

There were some who didn't approve of all the new projects. One of the more controversial improvements was the Bayliff Bypass. At least once a year for years, Tim Bayliff, the third generation at Chilancoh Ranch, wrote a polite letter to whatever minister was in charge asking that the road go elsewhere than through his front yard. He had a point. The road was barely spitting distance from the main ranch house (not that any of the Bayliffs spat), and it made a vicious turn almost onto the back porch. The dust and noise were awful and there was one serious accident. Fraser was sympathetic, and in 1978 he selected a line that would bypass the river road and avoid landslides and archae-ological sites. The bypass leaves the old road just past Bull Canyon and clambers up and over the hills for eleven miles before drop-ping down again just east of the Redstone bridge. Almost every-one except the Bayliffs and the Bill Blisses groused about the bypass. Truckers said the grade was more than the 8 percent Highways claimed it was. Tourism promoters and infrequent travellers were mad because the bypass misses the magnificent scenery along the river. Everyone thought it cost too much ($1 million), had too many cuts and fills, and should have gone a different way. Sightseers persisted in using the old route. Fraser defended the bypass, explaining it had to go so far off into the bush to "help the agricultural side of things." Truckers called it Fraser's Revenge, but they got used to it.

Landmarks disappeared along with the bumps and twists. The new road needed the land occupied by Gan Gan Lee's old log store at Lee's Corner. A Williams Lake man, Doug Shaw, salvaged the building. The new road also needed the land occupied by Tommy Lee's store at Alexis Creek, but no one rescued it. Stuart's at Redstone was bypassed, ending forty years of dodging Harold's bits and pieces. The McClinchy/Dowling Ranch (then known as the Dane Ranch) was also bypassed, a major diversion that took the road over the southern hill, forever out of reach of the flooding Kleena Kleene River. Evan Jones suggested that route fifty years earlier. The new road fixed the crossing at Brink's (it doesn't flood any more) and McClinchy Hill, Caribou Flats and all the other places that plagued motorists for years.

The Alex era was not without its trauma. When the Drummer gets together with the snow gods on the mountain, it still takes everything road crews have to keep it open. In the winter of 1975–76 a particularly heavy snowfall challenged the modern equipment and nearly won. In a twenty-four-hour period, five feet of snow fell on top of forty inches that hadn't been ploughed. It kept snowing, and the crew couldn't cope with it. There was fourteen feet of snow at the summit and thirty-nine avalanches. The first fall came so early some game guides couldn't get their

*Riske Creek store, 1989. Bert Roberts built a store, cafe and cabins just up the road from Becher House. It became the centre of Riske Creek after Becher House faded. The place has changed hands several times, and this is a new building. The original buildings are behind it.*

*Rock slide. This giant boulder came down on a sandy stretch of road just above Stuie, near Sand Hill, in the mid-1970s. Pictured with it is former highways foreman Ron Sollid. The son of Bella Coola pioneers, Sollid is with Caribou Road Services, the private company that maintains Highway 20.*

*Towdystan, 1990s. Norwegian colonist Jakob Lunaas settled at Towdystan around the turn of the century. For years his grandson, Fred Engebretson, was the sole resident of Towdystan, which is actually the Lunaas/Engebretson Ranch. Towdystan is one place that hasn't changed much over the years.*

*Redstone cemetery. The last resting place for members of the Alexis Creek band, the cemetery is right beside the road. A statue of the Virgin Mary watches over the white-fenced graves, which rest on a gentle slope.*

*Around the bend. Don Widdis pioneered trucking on the Bella Coola connection with the Hodgson Brothers' freight trucks. He later went into business on his own. As this picture shows, longer trucks had trouble wiggling around the hairpin corners and had to back up a few times to make it.*

*The new Highway 20. The road now takes logging trucks in its stride, but it wasn't always so. Even after the old Sheep Creek Bridge was replaced in 1962, there was still Sheep Creek Hill, windy and treacherous. This is a Timberland Holdings truck hauling in the Tatla Lake area.*

horses out and had to shoot them. Helicopters checked for stranded motorists and hauled fuel to the bulldozers. It was usual for heavy snow to bend the smaller trees over to just the right height to whack grader windshields, but this snow bent the trees into humps resembling igloos. One hump turned out to be the camp of some American hunters who threatened to sue the government for stranding them. They'd run out of cigarettes.

Bulldozers were brought in from everywhere. The road crews worked twelve-hour shifts with no days off for weeks—it was just like the old days, but the pay was better. Seven big bulldozers and a snow blower working full time couldn't keep ahead of it. One hired D8 worked twenty-four hours a day every day from November to January.

By 1980 the pavement had reached the top of Lee's Hill, and a road maintenance depot was established at Tatla Lake. People thought the depot was the forerunner of some diabolical plan to privatize the road crews, but if it was, Alex certainly changed his mind. Six years later when Premier Bill Vander Zalm did privatize highways, Alex fought it tooth and nail. Still, the Tatla Lake operation was different. It consisted of a small office trailer and a foreman who was allowed to park his house trailer on the site. He had a government pickup, but the road work was done by local equipment owners hired on a day labour basis. Don Piper was the first, last and only Tatla Lake foreman. A seasoned soldier of the Alexis Creek and Williams Lake crews, Piper had a thick skin and he needed it. The armchair experts were forever on his back. They complained—sometimes to Alex personally—when Don as much as farted. He was criticized for work he did, work he didn't do, for having a crew out when he shouldn't, and vice versa. Once he didn't have any crew to send.

There was an ice storm at the end of the Christmas holidays, just when everyone was coming or going home. Two inches of solid ice covered the road between Puntzi and the McClinchy and nothing could stay on it. Piper called out the two local dump trucks to sand the road, but the rear end went out of one before it got going. The other one made two loads, slipped on the ice and rolled over. The Alexis Creek and Anahim crews sent trucks to help but there wasn't much they could do because it rained and then froze, sandwiching the sand between layers of ice. Vehicles were ditched all over the place. Tatla Lake was on a community telephone line. People kept calling Piper's, mostly late at night, to find out what was going on, and the constantly ringing phones woke up everyone on the line. People were glad when the weather changed.

Don's wife Ann came in for her share of poor press too. One cold rainy day she came home to find one of the newer residents waiting in his pickup. He proceeded to harangue her about "his" road, threatening at one point to have Don fired. Ann was wearing a heavy-knit sweater that got heavier by the minute as it soaked up the rain. Water was dripping down her neck and nose and she was drenched by the time his record ran down. As he left, he reached out the window, patted her soggy shoulder and said, "Now you-all have a good day, you hear."

By 1986 the road had been rebuilt and paved to Chilanko Forks, and was "good gravel" from there to Anahim. The Bella Coola Hill was wider (but not lower), and the mountain road had been widened, straightened, and built up. East Branch was by-passed. The valley road was paved all the way. About the time the Social Credit Party was choosing a new leader, Alex was diagnosed as having throat cancer, and his larynx was removed. A number of operations later he was rigged out with a voice box. Although Alex was not a Vander Zalm supporter, the new premier indicated he would be highways minister as long as he wanted to be. Alex went into the following election campaign with all flags flying and his wife Gertrude at his side to be his voice. Cariboo was a dual riding for the first and only time, and Alex not only won his own seat by 6000 votes, he carried the second Socred, Neil Vant, along on his coattails. When the election dust settled, Vander Zalm dumped Alex and named Vant as Minister of Highways.

Alex was seventy years old and he was ill. He had served long enough to get his pension and could have retired with honour, but he didn't. He stayed to the end, fighting Vander Zalm over the highways privatization issue as ferociously as he had once fought the NDP. The King of the Cariboo died in the spring of 1989. He had a tremendous funeral. And he had his revenge, in a way. When a by-election was held that fall to fill his seat, Cariboo astonished itself by electing an NDP MLA. Cariboo born and bred rancher David Zirnhelt won by 6000 votes over the Social Credit candidate. Few people thought it was a coincidence.

A short time later, Vant was relieved of the highways ministry, but by then the Chilcotin Road was paved between Towdystan and Anahim Lake. This is disconcerting to longtime travellers, who expect the road to deteriorate as it goes west. It always did. Instead of good, fair, bad, worse, it is now blacktop, gravel, blacktop, gravel again over the mountain, then blacktop in the valley. The new Chilcotin Road makes it possible to drive from Williams Lake to Anahim safely in less than four hours, including pit stops. In the 1960s the trip took over seven hours, all being well, and in the 1950s ten hours. Before that the trip took several days. Now a traveller can get to Bella Coola in six hours without hurrying, though not necessarily in winter.

# CHAPTER 25

# Sundown on the Last Frontier

*There was smog over Anahim Lake, from the sawmill. What would Lester have said about that?*
—Steve Dorsey

Indeed, what would Lester have said. Or Andy Holte. They were gone before it happened, but Fred Engebretson saw the smog. He wasn't impressed, but then he was worrying about the environment long before it was the popular thing to do. "This land will provide for you if you don't tamper with it too much," he used to say. The road still runs west, into the sunset, but it is a hurrying road now. Travellers don't have time to stop for a visit, or a drink, or to listen to the Drummer. Just as the gravel overtook the dirt road, and the dirt road overtook the trails, the grey ribbon of asphalt is overtaking the gravel. With it, progress is overtaking the last frontier.

Chilcotin guarded her resources well. Her climate and vast distances held progress at bay. Over the years people prospected for minerals, and a few mines were developed, but nothing much happened, at least nothing that lasted long. In the 1970s an oil company eyed the plateau, leaving a grid of trails all over the place. Nothing came of that either. Only when the loggers brought in their monster machines that gobble whole forests in a day was it apparent Chilcotin had met her match.

Until the 1980s, the spindly lodgepole pine that freckle the plateau were considered more or less a weed, useful for fence rails and firewood. Nobody cared about them until the mountain pine beetle attacked them, then everybody started getting upset. The beetle lays its eggs under the bark of a tree, girdling the trunk. When the eggs hatch, they eat the wood. The tree eventually turns red and dies. Old-timers like Fred Engebretson, who believed a good fire once in a while cleansed the land and made the grass grow, blamed the forest service for the beetle. For years fire-fighters doused every little blaze, even the ones ranchers set in the spring to burn the old grass on their own meadows. Fred thought they overdid it, and the beetle epidemic was the result. Others blamed the warm winters. They said a good old-fashioned "50 below" winter would do the trick. Experts agreed, but the weather didn't co-operate and no cold winter materialized. No one knew what to do with the rapidly spreading red forest (old-timers said just leave it alone, but nobody wanted to do that), until the provincial forest ministry decided to log it. It is ironic that for so many years Chilcotin bugs kept people out, but the beetle brought them in.

Suddenly, Chilcotin's best weapons weren't good enough any more. Distances shrank with the improved road; hydro power de-clawed winter; the lumber industry had such sophisticated tools it was both practical and profitable to log even the remote areas. Before long, machines were chomping their way through stands of jackpine like ravenous dinosaurs, and there was even a sawmill—and smog—at Anahim Lake.

The Tsilhqot'ins, and then the settlers, made their living from the land as it was. They hardly left a mark. True, the main road cut across the plateau, and side roads trickled out of it, but every road had a destination, minded its business, and went only where it had to. Logging roads pry and poke their fingers into Chilcotin's private parts. Clearcuts leave her bare and defence-less—even the mosquitoes have no place to be. The spoor of civilization is everywhere, beer can rings and filter cigarette butts litter the lakeshores, used disposable diapers and plastic bags decorate the ditches. Some old-timers, and some newcomers, are fearful about the future. They say the high land heals slowly and they wonder if it can survive. They wonder if the wild animals will survive. They wonder if Chilcotin will survive. They hear the songs of the wilderness giving way to the thunder of progress and they don't like it.

The road still runs west, but most of the landmarks are gone, bypassed by the roadbuilders or stolen by time. Like the bleached bones of the russell fences that lie behind the Ministry of Highways utilitarian wire, a few ghosts of the past linger here and there. The old Sheep Creek Hill road hides behind the bushes, shadowing the new road. The historic Hilltop ranch house at the top of Sheep Creek Hill is still there. Johnny Wall's little log cabin has survived the battering of time and weather and sits, empty-eyed and lonely, on a rise overlooking Becher's Prairie. Johnny Wall was a harness maker, and the stretch of road by his cabin was called Wall Street.

Becher's Prairie has been tamed; some say the wind doesn't blow so hard across it any more. The first to be settled, Riske Creek was the first to go modern. There are still Jaspers at Milk Ranch, and the old barn is still there. The Lodge is still operating, but the main landmark now is the soaring Loran C tower that safeguards ships at sea. The Riske Creek store is in the same location but it is in a new building—painted bright canary yellow to match the highway's centre line. Becher's Dam hasn't changed much. The old trees in Hance's Timber have been logged. Old-timers insist those old pines never grew an inch. You can see the mountains now, and the land is open around Sawmill Creek. The Lookout at the top of Lee's Hill has been given its proper name, Hance's Lookout. Lee's Hill has been gentled, the Lee place has been split, but although there are "new" owners, the store and cafe are pure Chilcotin. Lee's Corner is now called Hanceville, because the post office is there. The old house at the Gables (Benny Franklin's first place) burned, but the pioneer connection hasn't broken—Norman Lee's grandson and namesake lives there. Chief Anaham's village still perches on the hill overlooking the valley which is even more productive now with its modern sprinkler system. The big Beware Rock (someone once spray-painted "Beware" on it) survived the roadbuilders.

Downtown Alexis Creek held its own against change for a long time. The road still squeezes between C1 Ranch buildings. Tom Lee's store was razed to make room for the road, but the lodge and house and the old police station are there on the high side. Jack and June Bliss converted the old police station into their home, the jail is now their dining room. The original hospital is still there, now operated by the Red Cross. Jakel's hotel burned

*Pigeon's Store, 1989. Gus Jakel started the second store in downtown Alexis Creek, but ever since he sold it to Eddie Pigeon it's been known as Pigeon's Store. Until the early 1960s, the top floor of the store served as a community hall, the scene for weddings, dances, church services and weekly movies. The store operated until 1993 when it was replaced by a new log building just around the corner.*

*The street where he lived. Peter Yells may be the only Chilcotin road foreman to have anything named for him. This is the "street"–more properly a stretch of road–where he lived in Alexis Creek, and that's Pete posing with the sign.*

*Red Hill, 1970s. When people talk about The Hill on the Bella Coola Road they mean the switchbacks. But Red Hill can be a horror when it's wet. The clay roadbed turns to goop at the slightest hint of moisture, and vehicles tend to slide off it. This picture shows Widdis driver Gary Tyrell managing to slide into the ditch instead of over the edge. It took a bulldozer and a grader to get him on his way again. Neither Widdis nor Tyrell remembers making one trouble-free trip.*

*Stake and rider fence. Chilcotin pioneers built a variety of fences, depending on the materials available. The log or snake fence required only some good-sized logs, an axe or saw, and muscle power. The picturesque Russell fences used smaller trees four to six inches in diameter, and lots of wire. (Old-timers claim they should be called hustle fences because you had to hustle to keep them in good repair.) Some people believe the stake and rider fence, pictured here, is the best for discouraging high-jumping animals. In Chilcotin it is unique to the Anahim Lake area.*

and was replaced. Pigeon's store is empty, replaced by a fancy new log building around the corner. Farther along, Alex Graham's second ranch is operated by a grandson, Norman Telford.

The big change comes after Bull Canyon (the early settlers fenced the ends of the canyon to pen their bulls in) just beyond Battle Bluff (battling Tsilhqot'ins pushed enemies off this towering cliff). Here the Bayliff Bypass sweeps up the hill and away from the river. The river changes colour, back to brown, farther upstream. It gets its distinctive milky turquoise colour when it is joined by the Chilko, near Bayliff's. You can see Stuart's abandoned store and garage through the screen of scrubby jackpines. The Redstone cemetery, its fenced graves the final resting places for members of the Alexis Creek Band, lies close to the road, just before you reach Redstone Reserve. Chilanko Forks has moved around over the years and now it is a one-stop service centre with a coffee shop and old-fashioned Chilcotin store. The old Knoll place is mostly underwater, a Ducks Unlimited project. Somebody cleaned up Gentleman Jim's parking lot, a graveyard of deceased autos that rusted quietly for many years beside the road near Pyper Lake. It's odd, but there is no geography named for pioneers like the Grahams, but a lake and a road are named for Bob Pyper, who never did much for anyone.

The remains of Tachadolier's log cabin still nestle in the bushes across the road from the meadow that bears his name, and is always spelled incorrectly on road maps. The cabin collapsed years ago. The logs are the colour of boiled beef. Tachadolier was a Tsilhqot'in, a loner by all accounts. The corduroy job done in the lane between the Tatla Lake Ranch meadows by Pete Yells's crew in the 1960s held up, but the only visible memento of the Graham clan is the Big House, now the Graham Inn. The downstairs has been rearranged and the parlour is now part of the restaurant but it still doesn't let anyone go away hungry. Tatla Lake hummed for a while with Forestry and Highways depots and it still is the area's nerve centre, with the store, post office and motel. The school is up the road a bit. The mountains start watching the road at Tatla.

The Bella Coola townsite and Williams Lake haven't moved,

so Tom Chignell's Half Way Ranch is still halfway between them. You can see the old McClinchy/Dowling ranch from the Dane Diversion if you can find an open spot between the trees. The Kleena Kleene River snakes its way through the ranch but it's hard to see it from the new highway. The Brink Ranch at Kleena Kleene is one of the breathtakingly beautiful spots in Chilcotin. Brink's is officially Kleena Kleene (the dot on the map); Leona Brink (Mrs. Fred) has had the post office and the store there for nearly thirty years. Clearwater Lake is now resorts. Baptiste Dester's barn and buildings haven't budged, they are empty but family members live nearby. McClinchy Hill is hardly a hill any more: the road on either end is straight and wide. No more rocks, no more slime. There isn't even one good mud hole left on Caribou Flats, and the road is paved from Towdystan to Anahim. Towdystan never changes, even with Fred gone. It's possible to whip right by Fish Trap without knowing it, but the remains of the earthworks dug by McDonald's crew in 1864 are still visible. It's marked as an historic site.

The community of Nimpo Lake is new in Chilcotin time. The service strip along the road had its beginnings in the late 1950s. Anahim has blacktop, hydro, dial telephones and modern buildings, but it is still the last frontier's last frontier. One memento of earlier days is an outhouse that sits under the trees. Ike Sing built it on the wrong side of his property line but nobody realized it until years later. When the other buildings were torn down and replaced, nobody bothered about the biffy. Christensen's store building is new (the old one burned) and boasts butchers and a liquor store. D'Arcy Christensen, Andy and Dorothy's son, was known as the Flying Fur Buyer. He flew his own plane into remote spots to buy fur rather than having the trappers come to him.

Few old-timers are left at Anahim. Most retire to warmer climates. There is a new hall at the Stampede grounds, which also boasts bleachers. The original hall collapsed under a deep snow but the new one isn't much bigger. When there's a crowd, dancers spend more time hopping up and down in place than they do moving around the hall. The hall has electricity but no plumbing. The campground has quadrupled in size but facilities are few and far between. There are two graveyards at Anahim. The Ulkatcho cemetery is on the reserve, the older graves covered with little houses. The pioneer cemetery overlooks the stampede grounds.

The Bella Coola connection has been tamed. East Branch has been bypassed (you can still get to it), but the Young Creek campsite hasn't changed. Even the mosquitoes seem to be the same. If you keep an eye out, you can still see bits of trees sticking out from under the original road fill.

The connection holds a few thrills yet. When the Drummer and the mountain gods get together they can still stop traffic, and for some reason, they have turned mean. Winter conditions caused no deaths on the road until the late 1980s, when a motorist was caught in the snow. She and her child perished before anyone knew they were missing. And the Hill claimed its first victims in the early summer of 1994. Three young west Chilcotin residents lost their lives when the pickup truck they were in went off the Hill near the bottom, and nothing broke its fall to the rocks below. A fourth passenger managed to jump out before it went over. The tragedy was doubly shocking because after forty years, no one thought of the Hill as a killer.

Bella Coola, like Anahim, has modernized physically but the spirit hasn't changed, neither have the surnames of many of the residents. The Native and white communities have lived side by side in Bella Coola—perhaps not quite together, but side by side—since the 1860s.

Everyone along the entire length of Highway 20 has electricity and telephones. Satellite TV dishes dot most yards, but some trappings of civilization are still missing. There isn't one supermarket or fast food chain outlet west of Williams Lake. The garages are locally owned. Travellers are unlikely to find a gas station or coffee shop open late at night and the restrooms at the Riske Creek Mega Fuels, and at all the rest stops, are of the outdoor variety. Chilcotin stores still carry an eclectic range of goods. Christensen's store boasts "If we don't have it, you don't need it." Most stores sell buckskin goods but now videotapes and scratch-and-win tickets jostle for space with black rubber gumboots and bolts of cloth. Most stores have a liquor outlet and the restaurants are licensed. Not everyone sees that as progress.

Chilcotin still has ranches, of course, some of them huge, and cowboys. Nothing has quite taken the place of a person on a horse to move cattle, although Ultra-Lite aircraft, all-terrain vehicles and 4x4 pickup trucks have taken over many cowboy tasks. Now the roads are good all year round, people think they need four-wheel-drive vehicles. It's hard to know who the cowboys are any more—logging truck drivers and Williams Lake lawyers are more likely to be wearing cowboy boots than cowboys are. Many "real" cowboys wear running shoes and baseball caps.

The road west, the third outlet to the sea, never did bring the prosperity the people who worked for it expected. Bella Coola is still waiting to become a major port. Chilcotin is still sparsely populated. The affair with the lumbermen is the best there has been in terms of prosperity. Mining remains a possibility. There is always ranching, but old-timers have their doubts about that. They say there are so many government agencies butting in with so many rules nobody can be free to ranch any more.

But some things never change. There still is a lot of grass and space in Chilcotin. There is still lots of elbow room. The Chilcotin sky is still higher and bluer than anywhere else. The air is still pure. In many places, the jackpine trees are still as thick as fur on a bear's back. The creeks and streams and lakes are still as clear as gin. The mountains and the hills endure. In a good year, the grass in the swamp meadows still grows hip-high to a tall horse. Winters don't seem to be as long any more, but spring still comes late in the high country. Summer leaves early; fall is magnificent.

The Drummer's beat is getting fainter, but you can hear it if you listen. You can hear it in the loon cry, in the coyote yowl, in the cricket chorus. You can hear it in the rustle of the cottonwoods by the river, and in the wind when it ghosts through the pines or howls across the flats. The Drummer is still there for those who want to hear him.

# Acknowledgements and Sources

Without the time and interest of the many people who shared their memories of Chilcotin, Bella Coola and the road with me, this book would never have been written. The interviews took place over a number of years, and sadly, a number of people have passed on. I hope they will live in these pages.

Special thanks for encouragement and advice to Elizabeth and Quint Robertson, and Wilna Wellingbrink.

Thanks to Dave Falconer; to local historians Dr. John Roberts and Irene Stangoe; to Irene Bliss, Edna Telford, Shirley Ross and Ruth Gurr for sharing their treasure troves of newspaper clippings, manuscripts and documents; to Tim and Merle Bayliff, Stan Dowling, Edwin Gaarden, Joy Graham, Vera Hance, the Hodgson family, Phil Robertson, the Tyrell family, Edna Telford, and the late Bud Barlowe and Mike Christensen, and especially June Bliss, who were so generous with their time and photographs.

Jim Raven, Frank Clapp, Richard Hadden, Cliff Parker and other British Columbia Ministry of Highways staff provided information or steered me in the right direction to find it.

Thanks also to the Canada Council Explorations Program for a grant which assisted in research.

### Interviews

*Interviewed by Glen Burrill:*
Dorothy Christensen

*Interviewed by Diana French:*

| | |
|---|---|
| Bud and Blanche Barlowe | Bill and Irene Bliss |
| Don and Marilyn Baxter | Jack and June Bliss |
| Tim Bayliff and Merle Bayliff | Frank Blunden |
| John and Gladys Blatchford | Fred and Leona Brink |

Victor Brink
Alfred Bryant
Vivien Cahoose
Tom and Eve Chignell
Mike Christensen
Frank Clapp
Vic Cockford
Gwen Colwell
Steve and Pat Dorsey
Stan Dowling
Bob and June Draney
George Draney
Jack Durrell
Isobel Edwards
Fred Engebretson
Harold and Alice Engebretson
Art Evans
Alex Fraser
Edwin Gaarden
Marge Gillis
Joy Graham
Larry Greenall
Dean Gurr

Mack and Ruth Gurr
Melvin Gurr
Dorothy Haines
Vera Hance
Elgin Heath
Wilf and Dru Hodgson
Gus Jakel
Ollie and Mary Knoll
Tommy Lee
Gordon Levelton
Clarence Mackill
Earl and Thelma McInroy
Don and Ann Piper
Phil and Joyce Robertson
Ike Sing
Thomas Squinas
Harold and Marcella Stuart
Bill and Edna Telford
Bill Thatcher
T. A. Walker
Don Widdis
Peter and Val Yells

*From Orchard Tapes, British Columbia Archives and Records Service:*
Mickey Dorsey and Eve Chignell
Rene Hance
Albert Franklin

*From Museum of the Cariboo Chilcotin (1967):*

Gay Bayliff
Bill Bliss
Andy Christensen
Dick Church
Lester Dorsey

Albert Franklin
Bob French
Dude Lavington
Tom Lee
Kathleen Telford

## Books

Akrigg, G. P. V. and Helen B. Akrigg. *British Columbia Chronicle 1778–1846*. Vancouver: Discovery Press, 1975.

_____. *British Columbia Chronicle 1847–1871*. Vancouver: Discovery Press, 1977.

_____. *British Columbia Place Names*. Victoria: Sono Nis, 1986.

Bancroft, Hubert Howe. *History of British Columbia 1792–1887*. San Francisco: The History Company, 1887.

Beeson, Edith. *Dunlevey: From the Diaries of Alex P. McInnes*. Lillooet BC: Lillooet Publishers, 1971.

Berton, Pierre. *The Last Spike*. Toronto: McClelland and Stewart, 1970.

_____. *The National Dream*. Toronto: McClelland and Stewart, 1971.

Boas, Franz. *The Ethnography of Franz Boas*. Edited and compiled by Ronald P. Rohner. Chicago: University of Chicago Press, 1969.

British Columbia, Ministry of Highways. *Frontier to Freeway*. Victoria BC: Queen's Printer, 1980.

Brown, R. C. Lundin. *Klatsassan and Other Reminiscences of Missionary Life in British Columbia*. London: Society for Promoting Christian Knowledge, 1873.

Centennial Project. *History and Legends of the Chilcotin*. Williams Lake BC: Cariboo Press, 1958.

Clemsen, Donovan. *Back Road Adventures*. Saanichton BC: Hancock House, 1976.

Cole, Douglas and Bradley Locker, eds. *The Journals of George M. Dawson 1875–1878*. Vancouver: UBC Press, 1989.

Collier, Eric. *Three Against the Wilderness*. London: Hutchison & Co., 1960.

Duff, Wilson. *The Indian History of British Columbia,* Vol. 1. Victoria BC: BC Provincial Museum, 1965.

Edwards, Isabel. *Ruffles on My Longjohns*. North Vancouver BC: Hancock House, 1981.

Farrand, Livingston. *Traditions of the Chilcotin Indians*, Vol. 2, Part 1 of *The Jesup North Pacific Expedition*. New York: American Museum of Natural History, 1900.

Hazlitt, William Carew. The Great Gold Fields of Cariboo. West Vancouver BC: Klanak, 1974.

Howay, F. W. *The Bute Inlet Massacre and the Chilcotin War*, Vol. 2 of *British Columbia from the Earliest Times to the Present*, ed. E. O. S.

Scholefield. Chicago: S. J. Clarke, 1914.

Hobson, Richmond P., Jr. *Grass Beyond the Mountains*. Philadelphia: Lippincott, 1951.

_____. *Nothing Too Good for a Cowboy*. Philadelphia: Lippincott, 1955.

_____. *The Rancher Takes a Wife*. Toronto: McClelland and Stewart, 1983.

Hutchison, Bruce. *The Fraser*. Toronto: Clarke Irwin, 1950.

Jackman, S. W. *Portraits of the Premiers*. Sidney BC: Gray's Publishing, 1969.

Kopas, Cliff. *Bella Coola*. Vancouver: Mitchell Press, 1970.

_____. *Packhorses to the Pacific*. Sidney BC: Gray's Publishing, 1976.

Lamb, W. Kaye, ed. *Letters and Journals of Simon Fraser, 1806–1808*. Toronto: Macmillan, 1960.

_____. *The Journals and Letters of Sir Alexander Mackenzie*. Cambridge: Cambridge University Press, 1970.

Lavington, H. Dude. *Born to be Hung*. Victoria: Sono Nis, 1983.

_____. *Nine Lives of a Cowboy*. Victoria: Sono Nis, 1982.

Lee, Norman. *Klondike Cattle Drive*. Vancouver: Mitchell Press, 1960.

Liddell, Ken. *This Is British Columbia*. Toronto: Ryerson Press, 1958.

Lyons, Cicely. *Salmon: Our Heritage*. Vancouver: Mitchell Press, 1969.

McIlwraith, Thomas Forsyth. *The Bella Coola Indians*. Toronto: University of Toronto Press, 1948.

Mackenzie, Alexander. *Voyages from Montreal on the River St. Laurence through the Continent of North America to the Frozen and Pacific Oceans*. Edmonton: Hurtig, 1971.

Morice, Adrien. *The History of the Northern Interior of British Columbia*. Fairfield WA: Ye Galleon Press, 1971.

Neering, Rosemary. *Down the Road*. Vancouver: Whitecap Books, 1991.

Old Age Pensioner's Organization, Branch #77. *A Tribute to the Past*. Quesnel BC: Spartan Printing & Advertising Ltd., 1985.

Ormsby, Margaret A. *British Columbia: A History*. Toronto: Macmillan, 1958.

Reid, R. L. *Pathfinders and Roadbuilders*. Victoria: Government of B.C., Department of Public Works, 1938.

Roger, Gertrude Minor. *Lady Rancher*. Saanichton BC: Hancock House, 1979.

Rothenburger, Mel. *The Chilcotin War*. Langley BC: Mr. Paperback, 1978.

Teit, James Alexander. *The Shuswap*. New York: AMS Press, 1975.

Usukawa, Saeko, et al, eds. *Sound Heritage: Voices from British Columbia*. Vancouver: Douglas & McIntyre, 1984.

Wade, Mark S. *The Cariboo Road*, Victoria: The Haunted Bookshop, 1979.

Walker, T. A. *Spatzizi*. Vancouver: Mitchell Press, 1970.

Whitehead, Margaret. *The Cariboo Mission: A History of the Oblates*. Victoria: Sono Nis, 1981.

**Newspapers and Periodicals**

*Ashcroft Journal*

*Bella Coola Courier*

*Bella Coola Mountain Courier*

*British Columbia Gazette*

*British Columbia Historical News*

*British Columbia Historical Quarterly*

*British Columbian* (New Westminster)

*British Columbia Road Runner* (Ministry of Highways)

*British Columbia Studies*

*Cariboo Digest*

*Cariboo Observer*

*Daily Colonist* (Victoria)

*Kamloops Sentinel*

*Kamloops Inland Sentinel*

*Northwest Digest*

*Victoria Times*

*Vancouver Sun*

*Vancouver Daily World*

*Province* (Vancouver)

*Williams Lake Tribune*

## Unpublished Surveyors' Notes

*British Columbia Archives and Records Service:*
  Poudrey, A. L.
  Swannell, Frank
*Collection T. A. Walker:*
  Lamarque, Ernest
  Rolston, Col.


## Other Unpublished Materials

The Clarke Collection, Museum of the Cariboo Chilcotin.

Goldman, Irving. "The Alkatcho Carrier of British Columbia." 1941.

Hall, John L. "Anahim Lake, British Columbia: A Study in
  Indian/White Interaction in a Western Canadian Ranching
  Community." PhD Thesis, University of Toronto, 1980.

Halverson, D. A. "Bella Coola: Local Level Politics." Special
  Collections, University of B.C. Library, 1973.

Laing, F. W. *Some Pioneers of the Cattle Industry*. Museum of the
  Cariboo Chilcotin.

Lane, Robert Brockstedt. "Cultural Relations of the Chilcotin Indians
  of West Central British Columbia." PhD Thesis, University of
  Washington, 1953.

LeBourdais, Louis (Cariboo politician and writer). Manuscripts and
  papers. Museum of the Cariboo Chilcotin.

Lee, E. P. Diaries, Cariboo Chilcotin Archives.

O'Reilly, Peter (land commissioner and magistrate). Letters; diary
  1866–1882. Museum of the Cariboo Chilcotin.

Palmer, Lt. H. S. Notes. Cariboo Chilcotin Archives, 1862.


## Government Records

Public Works Records. British Columbia Provincial Archives &
  Records Service, Victoria, BC.

Public Works/Ministry of Highways Records. Ministry of Highways,
  Victoria and Williams Lake, BC.

# Index